ELAINE KASKET

ALL THE GHOSTS IN THE MACHINE

The Digital Afterlife of Your Personal Data

ROBINSON

ROBINSON

First published in trade paperback in Great Britain in 2019 by Robinson

This paperback edition published in Great Britain in 2020 by Robinson

Copyright © Elaine Kasket, 2019

1 3 5 7 9 10 8 6 4 2

A CIP catalogue record for this book is
available from the British Library

ISBN: 978-1-47214-190-3

Typeset in Minion Pro by SX Composing DTP, Rayleigh, Essex
Printed and bound in Great Britain by Clays Ltd, Elcograf S.p.A.

Papers used by Robinson are from well-managed forests
and other responsible sources

Robinson
An imprint of
Little, Brown Book Group
Carmelite House
50 Victoria Embankment
London EC4Y 0DZ

An Hachette UK Company
www.hachette.co.uk

www.littlebrown.co.uk

For all those mortals who have
given me memories worth preserving

Contents

General Notes   ix

Introduction: Remembering Elizabeth   xi

Chapter One: The New Elysium   1

Chapter Two: The Anatomy of Online Grief   34

Chapter Three: Terms and Conditions   70

Chapter Four: Behind the Portcullis   102

Chapter Five: Stewarding the Online Dead   133

Chapter Six: The Uncanny Valley   167

Chapter Seven: The Voice of the Dead   203

Chapter Eight: Remembering Zoe   232

Last Words: A Decalogue for Your Digital Dust   242

Acknowledgements   252

Endnotes   254

Picture Credits   270

Index   271

General Notes

Interviews for this book, with a few exceptions, were audio recorded and transcribed in their entirety, with the permission of those being interviewed. All material contained within quotation marks consists of exact quotes, with ' . . . ' employed for omission of text within the same sentence and ' . . . .' employed for more extensive omission of material. In cases where audio recording was not done, interview material was transcribed in the moment, simultaneous with the conversation.

In cases where contributors' names and experiences were already in the public domain, by virtue of interviewees having participated in previous media interviews or academic publications, real names are retained. For situations where this was not the case, or when there was any doubt about whether living persons might be negatively affected by the inclusion of identifying information, names have been changed. I trust that where I have not undertaken direct interviews but have gleaned information from already published material, this is made clear within the text. I hope that all who contributed feel that they and their experiences were honoured and respected, even in those instances where I might have challenged some of their perspectives.

James and Elizabeth, about 1944

Introduction
Remembering Elizabeth

In the wake of my grandmother's death, her dutiful daughter helped my grandfather to sort through his wife's possessions, and to generally organise the house around his new life without her. As my mother encountered each room, drawer and cupboard, she was continually reminded that my grandmother saved nearly everything. This often had more to do with parsimony than sentimental attachment, for my grandparents both possessed a 'Depression mentality', a frugality forged in the American economic crisis of the 1930s. They sliced up expended cans of motor oil to make biscuit cutters. At Christmastime they saved whatever paper, string, ribbon and even tape was salvageable after presents had been unwrapped. Neatness and orderliness saved them from being hoarders, though, and, true to form, my grandmother's material legacy was tidy and often thoroughly documented with handwritten notes.

Sometimes Elizabeth's scraps of paper served to stake a claim, to assert her ownership of contested family items: 'This gold bracelet was one of a pair belonging to my aunt. Bernice wanted it, but it fell to me.' Sometimes they functioned as *aides-mémoire*, preserving details that for whatever reason Grandma deemed worthy of bequeathing to posterity. A cache of vintage sewing fabrics was accompanied by a comprehensive list of what they'd been used to create. Perhaps my grandmother intuited that these artefacts would be of some value or interest to her descendants, and she was correct in this. My mother used the information to pin notes about provenance to each surface of a set of cloth blocks that my grandmother had made for me in my infancy. 'Mandarin collar, slim skirt' says the note pinned to one side of a cube, a square of beige floral cotton. 'Scalloped hem, going away after wed[ding]' says another. One note, pinned on a print featuring whimsically cartoonish lions, reads 'my big dinner date in Toronto (camp trip dress)'. These keywords immediately linked the fabric to a tale I'd heard several times, about an unsuitable suitor chatting up my teenaged mother at a Canadian campground. I didn't remember playing with these blocks as a child, but when my mother dug them out to share with me, we fingered the textures of the materials and talked over the memories connected with each garment. Turning the blocks over and over on the floor and aiming the camera of my iPhone, I snapped photos of each side of each cube, complete with their pinned notes, shunting the images into a file in the cloud titled 'sewing room collages'. I intended to use them to decorate the wall next to my own sewing machine, 4,127 miles away from where the dresses and blocks had been made.

Toy fabric blocks made by Elizabeth, with her daughter's pinned notes.[1]

In my grandparents' house, my mother found one thing, however, that had escaped labelling. It was a medium-sized plain cardboard box, sitting anonymously at the bottom of the closet in a guest bedroom. Caught up in a torrent of efficiency, and no doubt assessing the box for potential disposal, my mother asked her father, James, about its contents. He didn't have to check. He didn't even have to think about it. 'Love letters', Grandpa said immediately, with an air of nonchalance. Then, according to her account, they went on with their clearing-out, leaving the box where it lay.

Listening to my mother describe her recollections of that day, two decades on from the event, I was overcome with curiosity. Did Grandpa give her any more details about these letters? Had she peeked into the box on the day, or while my grandfather was still alive? Did she get the sense at the time, especially given his swift response, that my grandfather had looked at them lately? Did he give any indication of what he wanted done with these letters in the future? No, no, no and no. It wasn't until my grandfather himself died, and my mother was preparing the house for sale, that she encountered the box again. My mother, in her seventh decade and newly orphaned, took the box home.

In the twilight years of Grandpa's life, he had taken to storying his sixty-year marriage to my grandmother in a particular way. 'I have loved Elizabeth my whole life,' he would say, 'but she never loved me.' Some members of the family found this statement to be entirely plausible, for they viewed my grandmother's character and her behaviour in a light situated at the cool rather than the warm end of the spectrum. She was universally agreed to be intelligent, strong, determined and courageous, but she brought a fierceness, a flintiness, a steeliness to all these qualities. Her children had witnessed the hardness with which she sometimes spoke to her husband, and her grandchildren had experienced the painful, hard angularity of her awkward, infrequent hugs. As adults, some of my uncles felt sorry for their self-professedly lovelorn father and allied themselves with his narrative.

This was Elizabeth the imperious snow queen, her affections withheld behind a wall of ice that the warmth of my grandfather's affections was insufficient to melt. Something in this characterisation, however, jarred with my mother. The assertions that her mother had never loved her father – so readily accepted at face value by some – cut her to the quick and didn't even seem accurate. She possessed, however, nothing concrete to show that this picture was anything other than true, that there was any more to Elizabeth than this. Although my artist mother was unhappy with what seemed to have become the official portrait of her mother, she lacked sufficient materials to paint anything different.

When my mother took the box home, that all changed. Lifting the lid, she opened not just the flaps of cardboard but a window on her parents' relationship. Reaching in, she encountered the first layers of an extraordinarily prolific correspondence between her mother and father, dated 1945. My grandfather had been dispatched to a training camp in preparation for fighting in a war that came to an end before he was able to board a ship. At that moment in time, my grandparents had been married for ten years and had three young children, including my mother and two of my uncles. When James was called up, it would be the couple's only extended separation since the first flush of their teenage romance. While he was away, it was not unusual for them to write to one another three times a day.

This archive my mother had unearthed proved fascinating on multiple levels. For any reader, it had value as a vivid portrayal of the domestic life of a working-class family at a critical juncture in American history. My grandparents wrote with both remarkable skill and exhaustive detail and shared a gift for observation and description. For my mother, however, it was much more than this. It was not the insights into mid-century wartime America that kept her transfixed in her chair for two weeks.

Elizabeth to James, May 1945. *My greatest hope is that this doesn't have to go on very long. Even if the war ends in six or three months it's not soon enough for me. The longer you are away the harder it is to take it and the more lonesome I get . . . . I've been awful lonesome for you today. It was quiet here and lonely and my tummy kind of hurt. Sometimes I wish I could go to sleep and not wake up until you come back . . . I spent 1 hour and 48 minutes writing this letter but you are worth it . . . . I really enjoy writing to you because I just feel like I'm talking to you when I write and you know in our whole history I never get tired of talking to you.'*

In this and so many other passages, there it was, in black and white. While love can be hard to define and even harder to quantify, sometimes you just know it when you see it – it announces itself too clearly to be mistaken for anything else. My grandfather had been dead wrong about his wife never caring for him. He had mis-remembered Elizabeth. In 1945, even after a decade of marriage and parenthood, she had been truly, madly and deeply in love with him. The letters made it impossible for my grandmother to be reduced to the qualities that had been most visible in her last years, and her hardness had been a state rather than a trait, or at least a facet rather than the whole. Seeing her words, perceiving this incontrovertible truth, it was not vindication that my mother felt. It was healing. She felt that the wholeness of Elizabeth, and the reality of her parents' mutual love, had been restored to her memory. Finally, she had a picture of her mother, and of her parents' marriage, that she could carry forward with less sadness, and more comfort and joy.

As the self-appointed steward of this legacy of love, my mother proceeded to spend weeks painstakingly arranging the correspond-ence into chronological order, slipping each precious letter into transparent polyurethane pockets and clipping the pages into large ring binders. She encountered glimpses of her childhood self along the way, in the form of letters and pictures she had sent to her faraway father, and she once again read the replies her father had sent, written in print rather than cursive, adapted for a six year old's eyes. When she finished, the letters filled five thick volumes. Wondering if this archive would provide solace to others, as it had for her, she offered the binders to her brothers and any other interested family members. Her elder brother demurred, and he died without ever seeing them. Her younger brother, only a toddler in 1945, took the first volume but said little about it for months. My mother chased him about it and, after some searching, my uncle produced it from where it was buried, possibly unread, underneath a stack of books and papers at the bottom of a cupboard.

My mother had hoped to share 'her' Elizabeth with others in the family, to use the archive to support her version of the truth: her mother as a multifaceted woman, in her softness and hardness, her vulnerability and strength. If she failed in her mission, it was not exclusively down to others' reluctance to engage with the letters. A unified, agreed-upon vision of Elizabeth would never have happened, even if each person had read the correspondence from beginning to end. Stage models of grief fail to adequately capture the infinite variety of individual bereavement experiences because the uniqueness of our relationships in life prevents a tidy, predictable, phase-by-phase unfolding of the grief process. Everyone in my family experienced and remembered a different Elizabeth, and so each person had their own requirements in grief.

Those letters did something for my mother. They salved her pain and rectified a painful dissonance in her mind, ultimately helping her arrive at a biography of her mother that she could comfortably carry forward. Maybe my uncles, the ones that didn't read the letters, didn't share my mother's particular hurt. Perhaps there was nothing for them to resolve. Whatever they had needed to carry on, maybe they had already found it, and because of that, the correspondence would have been disruptive, not comforting. When I asked my mother why her younger sibling had left the letters unread, despite his weakness for nostalgia and his fascination with history, her answer was telling. 'They made him too emotional,' she said. The same thing that had helped her find her way only served to throw her brother off course.

Although she ultimately accepted and respected others' reluctance to delve into the letters, I can understand my mother's abiding wish for some of the prevailing family narratives about Elizabeth to be challenged and debunked. I can see that it remains painful for her when the complexity of her all-too-human mother is effaced, when others reduce her to a monolithic ice sculpture. No wonder she was so pleased, then, when I took an interest in the volumes of wartime correspondence. Settling in to read them, and having no idea what to expect, I was shocked at their intimacy and warmth. Sometimes I caught my breath, and sometimes I blushed up to my forehead. On more than one occasion, I was moved to tears. Like my mother, I immediately felt compelled to share them with members of my own generation. I transcribed some passages into the notes app on my phone, and later that night read out some of the excerpts to my siblings and cousins, including something Elizabeth had written in May 1945.

If you were here, I'd love you and hug and kiss you enough for the next 20 years. You are always in my thoughts.

My cousin's eyebrows shot upwards in surprise. 'Wow,' she said. 'That doesn't sound like her at *all*.' Around the table there was consensus that, indeed, it didn't seem possible that this quote could come from a letter written by the Elizabeth we'd known. Someone – perhaps more than one of us – laid it on the line. *Frankly, she could be kind of a bitch.*

But her letters made it clear that things were more complicated than that and, like my mother, I felt an urge to explode the myth. Back at my parents' house, I sat in bed and scrolled through my typed transcriptions in the notes app on my iPhone. Again and again, my eye was drawn to one passage in particular. While no one could describe it as vulgar, nor could it ever read as chaste. Like poetry, it contained great depths within its brevity, suffused as it was with helpless desire, and it conveyed a passionate attachment that was as sexual as it was loving. It intimated the kind of emotion that most of us want to experience as much as possible in our lifetimes, and reading and re-reading it, I felt the frisson of a voyeur. Even so, I grabbed a Moleskine notebook and a favourite pen, and I reverse-translated my note from typescript back into longhand. Altering the dimensions on my phone's camera settings to square, I photographed my own writing, applied what seemed like an appropriately sepia-tone filter, and added a caption: 'True romance'.

Then, not without hesitation or compunction, but without anyone's permission but my own, I posted it on Instagram.[2]

"I think of you
every time I can,
and sometimes
when I'm not supposed to."
James to Elizabeth
April 1945

Ω

While the letters meant a lot to my mother and she hoped that others might read them, she stopped short of imposing her own meanings on anyone else. When she assembled those binders, she added nothing, made no comment, appended no annotations or interpretations. She also left nothing out. It's comprehensive, all there, arranged in chronological order: the letters, the children's drawings, the photos, the envelopes that had carried them. She made the archive more accessible and readable, but she didn't transform it. For better or worse, when I encountered them, I went about things completely differently. Where she stayed faithful, I translated. Where she compiled, I curated. Where she maintained privacy, I seized control of the border between public and private and shifted the boundary line.

In short, where my mother stayed analogue, I went digital. Holding up a digital prism to a static, material legacy, I turned the prism this way and that, watching the colours change, finding the refractions that I liked best. And I wasn't alone – I invited a crowd of onlookers to watch along with me. I took a private correspondence that was not intended for me and transformed it from its original functions: I wanted it to challenge others' ideas, and to affect their emotions, the way the passage had affected mine. I wanted others to associate it with me. Ladies and gentlemen of today, I present to you the deep love my grandparents held for one another in 1945. 'True romance', heart emoticon. But for this love, I wouldn't be here. It seems like a lovely thing to share. And yet, even as I write these words, I recognise that certain individuals might see my Instagram post as a transgression. Some of those people are alive, and some are not. In the latter category, of course, fall Elizabeth and James themselves. Fortunately or unfortunately, they're not around to pass comment or to click on the little heart below the post to turn it red, to notch up my 'likes'.

My grandparents never indicated, by word or deed, that this evidence of their love was to remain private, unread by others for evermore. Preserved intact, behind no lock and under no particular instruction, it was anyone's guess whether they had been left available to us by intention or by default, and everyone had their own take on it. My mother, compelled by her own desire and driven by her own search for solace and meaning, seemed to take it for granted that kinship conferred right of readership. My uncle's ex-wife, still heavily involved with the family but sometimes unsure about whether her kinship ties remained strong enough for her to consider herself inner circle, expressed concern about whether she should be reading the letters, conscious of their intimacy. My mother fielded my questions about privacy with a startled air, as though she were reflecting on it for the first time. It's not a surprise to me, however, that of all the themes woven throughout this story, privacy is the thread I pick out most readily. I don't value it so closely because its importance has been inculcated in me since childhood, because that isn't the case. It's because, as someone who is positioned at the junction where death, technology and psychology meet, I spend most of my time thinking about how the digital world affects us.

Every day that the sun rises on the citizens of the information age is another day that privacy is tossed into the searing-hot crucible of the digital age, where it is tested to its limits. Online, privacy assumes whole new levels of complexity, constantly shifting and morphing in its nature, always up for scrutiny and debate. For generations it was taken for granted in law and in commonly received wisdom that privacy, like other human rights, belonged to 'natural persons' – living persons. As the breath left their bodies, the dead relinquished their right to and indeed their need for privacy. Privacy is partly about the right to self-determination, and when you were dead, there was no more 'self'. But now, asking whatever experts you can rustle up, just try to work out a definitive answer to the question of whether the dead have the right to privacy. You might first need to establish what is meant by 'dead': are we talking dead in body, or socially dead? Once you establish that, you will likely still receive a dozen different explanations of why they do or don't have that right, all of which and none of which will be true.

Anyway, the question of whether I'll have a right to privacy when I'm dead may be a moot point. The way things stand, if I or my beloved go off to fight in a modern war and we have a beautiful, eloquent, impassioned correspondence, that correspondence will be conducted via email, WhatsApp, Facebook Messenger and mobile phone. Because my privacy settings on all of the above are configured like a maximum-security information prison from which no data can escape, my next of kin would have a snowball's chance in hell of ever laying eyes on even a fraction of it, despite what they might wish or need. How you feel about that state of affairs depends, at least in part, on whether you are a digital immigrant, or a digital native.[3]

However au fait you may be with digital technologies, if your formative years occurred well before the mid-1980s, you can consider yourself to be a digital immigrant. Being one of their number, I can sketch you the profile of someone in this category with intimate familiarity and no small amount of nostalgia. Digital immigrants remember keeping address books, writing the names in pen and the addresses in pencil, as their mothers suggested. They remember when a letter being returned as 'sender not known/no forwarding address' meant that you might never find that person again. They know what a mimeograph machine is and can recall the smell of the sheets still damp off the roller, the purple colour of the ink. They remember flicking through card catalogues in libraries, hitting 'record' on a tape deck when a favourite song came on the radio, and having no way to surreptitiously cheat at the pub quiz or at Trivial Pursuit. They recall the loneliness of being in a phone box far from home, perhaps in a foreign country, running out of change, hearing the operator's warning and the line going dead, not knowing when they might hear that familiar voice again. They recall going to the record shop to buy LPs, long before the digital-backlash resurgence of the turntable. They remember their wonder at the shiny-smooth space-age surface of their first CD or DVD. They can mimic the sound of a dial-up connection. They remember the suspense of collecting photos from the developer, and the disappointment, after such a wait, of discovering that the film was overexposed, or that there was an errant thumb in the important shot. And they tend to have albums or shoeboxes full of those photos, fading and curling snapshots, negatives going bad in the heat of the attic, mixed in with letters and other ephemera. Of the memorabilia that matter, a goodly percentage will be of the type that is vulnerable to mice and silverfish, not corrupted files and motherboard failures.

As savvy as I may become with digital technology, and as much as I might use it, I shall never be anything more than an interloper, someone who still experiences a frisson of excitement and awe during a video call, even though I Skype and FaceTime every day. I'll never be able to live in the skin of a digital native like my eight year old, who regards these phenomena as much a part of the furniture as the table on which she eats her breakfast. She grew up believing that WiFi is everywhere in the air, part of the molecular structure of oxygen, and perceiving it to be nearly as critical to life. She is neither grateful nor amazed that she can contact anyone anytime from anywhere, or that her access to the programmes she likes is instantaneous and within her control, or that the answer to any question she might think to ask can be answered by Google, Siri or Alexa.

On the contrary, it is so much her birthright that she cannot imagine it to be otherwise. Having absorbed via osmosis the social instincts of her digital age, she requests that photographs she takes be shared on Instagram or flogged on eBay and is deeply unimpressed at not being permitted to vlog about her hobbies on YouTube, like some of her little friends. Whether through vague notions of the importance of privacy or just preoccupation with what is 'hers', she has installed a passcode on her smartphone, just like Mummy. When she enters adolescence and has full command of her own devices, both her deliberately stored memories and the inadvertent traces of her life will likely be archived endlessly and voluminously every day, but probably almost exclusively in digital form. If even 1 per cent of my daughter's existing digital footprint were to survive, future generations, encountering it, could come to feel that they know her intimately, could sense her personality with no difficulty whatsoever, like falling off a log. I feel entirely confident in my assumption that she will never have a shoebox full of letters in the attic and, when it is her time to die, I don't know whether it will occur to any of her contemporaries to visit her grave. Why should they? She'll be in the palms of their hands.

Today's filtered photographs and unfiltered disclosures would have flummoxed my grandparents, who were born, lived and died before ever being presented with the opportunity to become digital immigrants. So, at first glance, the story of the two of them, their legacy and the different ways my family members have dealt with it, doesn't seem to have much to do with the digital age at all. But what do the themes of identity, loss, connection, memory, control, owner-ship, stewardship and privacy have in common? They all have to do as much with life as they do with death, and they're all deeply affected and challenged by the digital age. Not particularly gradually, not bit by bit, but suddenly and in a tsunami of bytes, everything about these concepts has changed. Whether you are a digital immigrant or a digital native, this book is designed to make you think not just about your death, but about your life. The existential philosophers I studied in graduate school would probably have predicted this, but it turns out that death in general, and death in the digital age *particularly*, is an uncannily useful vehicle for thinking about choices we make in life, considering what's important to us, and adjusting our actions accordingly.

This book, therefore, is about your relationships with the people whom you love and who love you, what you mean to each other and how you connect with one another, and how you would like to stay connected after you or they are gone. It's about how you want to be remembered after your breath becomes air,[4] and your body becomes earth, and how the digital age can make that happen in ways that were unimaginable even a decade ago. It's about how you decide what's private and what's public, now that the information age has turned historical expectations and definitions of privacy upside down, and about how the boxes you tick on Facebook's settings could decide how you are remembered for ever. It's about how the individuals or forces that own or control your data can ultimately determine not only what remains of you in the world, but also who can access what's left behind. It's about the desegregation of the living and the dead online, the places where they mix and meet and socialise, and about how one day, not too long from now, some of your best friends and most sparkling interlocutors could be dead. It's about our fantasies of eternal life, and the tantalising hope being held out by digital technologies that maybe, just maybe, we are beginning to figure out how to cheat death.

So, would you like your legacy to live on in the world for evermore, or would you like your digital traces to disappear altogether, swept away like a wave effacing your footprints from the sand? Do you prefer the idea of digital immortality, or both physical and technological obsolescence? Think carefully before you answer. And whatever you do, once you choose, have a *very* careful read of the terms and conditions.

Chapter One
The New Elysium

For ten minutes or so, I'd been hiding in the toilets. If I had ever felt more deflated and embarrassed, I couldn't remember when it had been. I'd just given this very same talk at a professional conference a few months before, and it had been a big success. Social media were a relatively new thing and mourning on social media was a really new thing, so I was one of the first people talking about it. There'd been so much good feedback from the academics in my field at that conference, so much robust discussion.

Today, however, was an entirely different kettle of fish. Nearly every person in attendance had looked checked-out or confused. I'd seen lots of crossed arms and other discouraging body language, and you could have heard a pin drop when it came time for Q&As. I'd clearly misjudged my audience, made up of members of the public this time. How could I have got it so wrong? I'd just plunged into my presentation assuming that everyone knew what social media were and how they worked, but the majority of them clearly had no idea what I was talking about for forty minutes. Unsurprising, really. In 2010 the uptake of social media amongst anyone over thirty was incredibly low. God, what an idiot. Splashing cold water on my burning cheeks, I gathered my resolve to go back out into the art gallery where the Stardust Symposium on death was taking place.[5] People were milling around, waiting for the discussion groups to start. Still feeling rattled, I took up a position in a corner and attempted, unsuccessfully, to fade into the wallpaper.

She saw me from across the room and made a beeline for me. Of all the digital immigrants in the room, she was surely the most senior. Wreathed in a dandelion puff of wispy white hair, she had an outsize, ancient handbag hooked over her arm, like the Queen. Although frail in gait and minuscule in stature, she was assertive in her approach and clearly had something to say, and I was pretty sure what it was. I smiled weakly and prepared myself for a Luddite diatribe about how these newfangled notions like the World Wide Whatchamacallit were a bunch of stuff and nonsense.

'You do realise,' she said briskly, 'that what you are talking about is nothing new. Nothing new at all.'

'Nothing new?' I said blankly. I had fully expected her to think that this was entirely too new for comfort. After all, Facebook had opened to the public four years before. Not only were people creating memorial pages on the site, but they were continuing to visit the 'in-life profiles' every day, repurposing them as sites not just for mourning, not just for memorialisation, but for *talking to* the dead. I had noticed that their talk was everyday, casual, nearly always phrased in the second person. 'I can't believe you won't be with us this year. God, that sucks. Miss you babe!!' Some comments seemed to imply that it was necessary, or at least more effective, to get online to speak to the dead. 'Sorry I couldn't say happy birthday yesterday – I couldn't get any Internet where I was staying.'

Sure enough, when I spoke to research participants, they confirmed that talking to someone on Facebook was the best way to ensure you were getting through. They didn't necessarily believe there was an Internet cafe in the sky. Not all of them believed in Heaven or some other form of afterlife. But religious folk and atheists said the same thing. If you talk to someone at the gravesite, or in their room, who knows if they hear you? If you write something to them on paper and leave it somewhere, who knows if they see it? But if you write to them on Facebook – oh yes, they'll read that.[6] How could this not be new? But it was so not-new, in fact, that she was practically bored. 'This is as old as the sun,' she said, sighing. She reached into the handbag and drew forth an envelope. Opening it, she handed me a stack of faded, rough-edged bits of paper, sepia-tone images of people in late nineteenth-century dress. The figures posed stiffly on their straight-backed chairs, their faces etched with sorrow mixed with stoicism. In the background, visible through clouds of soft mist, were transparent blurry faces and figures, ghosts of the dear departed.

'You see?' she said. She must have seen that I didn't see. 'I belong to the Glasgow Association of Spiritualists,' she continued. 'And this that you're talking about, it's just the same thing. We can always talk with the spirits. They find their ways.'

'These are spirit photographs,' I said, tentatively.

'Exactly,' she said, triumphantly.

I think she kept talking, but I don't remember what she said. By this time, my mind had snapped shut. I put my polite face on, my gaze flickering over her shoulder for a rescuer, someone I really must catch before they go. Clearly, I thought, this woman was a bit of a lunatic, bless her. She didn't understand this phenomenon or the point of my presentation. Wherever she is now, though, I'm sending out a message into the ether, perhaps the same ether the spirits travelled to show up in those photographs: I'm really sorry that I blanked you, madam. It wasn't that you didn't understand what I was saying. I didn't understand what *you* were saying. And you were making a fantastic point.

There is a theory of grief called continuing bonds.[7] While much more will be said in this book about continuing bonds, the essence is this: continuing a connection to the dead is an entirely normal phenomenon. If that sounds like a bizarre and even unhealthy idea, look around to see if Sigmund Freud's ghost is materialising in a mist behind you. Such was this early psychologist's influence on Western thinking that, temporarily at least, he was able to shift all of our existing assumptions about death and grief. His impact on our perceptions of 'healthy' grieving is all the more remarkable when you realise how little he actually said about it. In an essay entitled 'Mourning and Melancholia', he spelled it all out for us as being simply common sense. 'I do not think,' he said, 'there is anything far-fetched in presenting it in the following way.' He went on to explain that the 'work' of grief was all about a gradual, 'piecemeal' letting-go of the 'loved object' – the person who has died, that is. He was nonspecific about the details, but apparently you needed to sift through each memory and every lost hope that you cherished about the deceased person, following which you would be able to let go each of those bits – a steady, stepwise release from your pain. If the 'existence of the lost object is psychically prolonged', Freud said, we cannot be 'free and uninhibited', hence the necessity of 'working through' our grief.[8]

There was just one problem. These matter-of-fact pronouncements, imbued with authority as they were, didn't stand up to any scrutiny. They didn't map onto most people's experiences. They didn't even reflect Freud's personal experiences of loss. Neither empirical nor anecdotal evidence supported them, but the great man's version of where the dead belong in the grand scheme of things held fast throughout the remainder of the twentieth century, in the West at least. We hear his voice echoing down the decades when people employ phrases like 'not getting over it', 'finding it hard to let go' and 'being in denial'. Despite growing popular awareness that there's something amiss with these ideas, Westerners who experience particularly close bonds to their dead still tend to pathologise themselves, and to be pathologised by others. In actuality, they are merely responding to an urge that is as old as time – the cherishing and even nurturing of psychological and emotional ties with lost loved ones. For myriad reasons, many of which will be explored in these pages, staying connected to the dead matters to us.

Communication technologies provide a perfect illustration of how our urge to continue bonds with the dead seems to be wired into us, for no sooner does a new bit of kit emerge than we seize upon it as a way of getting hold of our dead. In 1840s New York, young Kate and Margaret Fox – clearly blessed with active imaginations and a keen sense of drama – managed to persuade everyone that they were in communication with spirits. Their careers took off like wildfire. In the intervening time between alleged first contact and the confession that it had been an elaborate hoax all along, the siblings enjoyed a lucrative few decades as mediums, responding to a seemingly limitless thirst for communion with the dead, and helping make the Spiritualist movement in the United States hugely popular. By the 1880s fake spiritualism was all the rage, the 'hot fraud' of the time.[9] For the Fox sisters' famous séances, the spirits' chosen manifestation is telling: while the miraculous new telegraph tapped, the spirits rapped. Whether an interest in the latest gadget directly inspired Kate and Margaret we shall never know but, the more that telegraphy spread, and people began to equate tapping with communication, the more sense it made for spirits to speak in the same way.

It didn't end there. As the popularity of photography increased and photographic equipment became more widespread and accessible, 'spook pictures' became a standard offering on the savvy spiritualist's menu. An investigative reporter presenting to Mrs L. Carter's Los Angeles photography studio in 1880 baulked at the stated price of $3.50 a sitting, despite being promised that she'd never had an out-and-out failure. If the spirit he was after was preoccupied with some other spirit-world business, she assured him, a stand-in would come along instead. When it looked as if the gentleman might take his business elsewhere, the medium received a message from the astral plane that a reduced fee of $2.50 would probably do.[10]

Unlike the enterprising Mrs Carter and her cynical customer, however, some people genuinely believed in the power of technology to capture manifestations of the spirits, including persons of such eminence as Thomas Edison himself. Having invented the phonograph in the late nineteenth century, he wrote in *Scientific American* in 1920 that he hoped one day to produce a phonograph sufficiently sensitive to pick up the voices of the dead.[11] For a population reeling at the sudden loss of so much young life in the War to End All Wars, this must have been an appealing prospect. By the 1980s, spectres were manifesting through the static buzz and flickering lines of untuned television sets in blockbuster films.[12] And in twenty-first century Japan, people journey from far and wide to enter the hilltop property of Itaru Sasaki and pick up the receiver of his wind telephone.

When Itaru's cousin died, the weight of loss was hard for him to bear. Particularly in the aftermath of a death, we often experience the urge to find and contact the lost person – the searching-and-calling reflex. Such was the case for Itaru, a gardener from Otsuchi, Japan, who happened to have a beautiful piece of land overlooking the Pacific. To satisfy his craving to connect, he installed a white-painted telephone booth in his garden. On the shelf inside he placed an old-fashioned black rotary telephone, which connected to nowhere at all. Itaru knew that actually plugging in this phone would be unnecessary for his purposes, for he didn't need to speak to the living, he wished to speak to the dead. 'Because my thoughts could not be relayed over a regular phone line,' he said, 'I wanted them to be carried on the wind . . . so I named it the wind telephone' – *kaze no denwa*.[13] Holding the receiver to his ear, gazing through the floor-to-ceiling panes of glass at his brilliantly coloured flowers blowing in the breeze, and at the blue of the sky and the distant shimmering of the sea, Itaru spoke his innermost thoughts and feelings to his cousin, comforted by his conviction that the ocean winds would bear them to their intended recipient.

Itaru built his telephone in 2010. In the following year, March of 2011, the unthinkable happened. Triggered by a massive earthquake, the Tōhoku tsunami roared ashore to sweep away thousands of lives, killing 10 per cent of the population of Itaru's town. In the years that followed, somehow word got out about his wind telephone. People began to visit his garden, first a trickle and then in numbers steadily mounting into the hundreds, and the visits continue today. Entering the booth, people dial the numbers for homes that were swept away, or mobile phones gone silent, the usual numbers on which they used to reach parents, husbands and wives, sisters and brothers, cousins and friends. Sometimes the visitors come alone, sometimes with their families, sometimes once and sometimes regularly. Sometimes they say nothing at all, and sometimes they give a full update on what is happening in their lives. Some pepper their conversation with questions and well wishes to the deceased; questions that they know will go unanswered, but that they ask anyway. When they come to speak to their dead on the wind telephone, they are in a place that their lost loved one likely never visited, and there is nothing but silence on the line when they lift the receiver of the wind telephone. There is no stamp of the familiar, no image of the dead person they have lost. And yet they come to use the phone.

The Fox sisters were hoaxers and the spook photographers were swindlers, and *Poltergeist* was Hollywood movie magic. The visitors to the *kaze no denwa* in Otsuchi may or may not literally believe that their communications reach their loved ones. Whether spirits actually exist or communicate across the great divide between life and death, though, is beside the point – people want to believe. Faced with a loss that seems impossible to bear, the most cynical amongst us willingly jumps at the chance of feeling a thread of connection again – whether they see that connection in literal or emotional terms – and we have long used technology to help us achieve that. On this level at least, the nonagenarian Glasgow Spiritualist was spot on. At the same time, it's not quite the case that there is nothing new under the sun. Technology used to be a *means* of getting hold of the dead, but it's no longer merely a medium to help us reach those who've gone before. Technology is where the dead live already.

This is the new Elysium:[14] not an exclusive place, not only for those favoured by the Greek gods, but for all of us. Provided, that is, you have a digital footprint. If you really want to be remembered, if you like the idea of people continuing to feel connected with you after you are no longer here, bear in mind that the size of that digital footprint is going to matter . . . a lot. More about that in a minute, but first let's consider something quite remarkable. So far, there are no digital worms, no virtual carrion beetles traversing the Internet, nibbling away all traces of dead people's data. A whole lot of information stays put – sometimes identifiable as originating from someone who's died, but often not. This state of affairs represents, at least, a bit of a throwback. At most, it may be the harbinger of a sea change in how we experience death, and in the place and influence of our ancestors in society. Let me explain.

Ω

We tend to speak about many Western societies as being in collective death-denial mode. Just listen to how people talk about talking about death. We don't talk about death in our culture, say the English, for example; it's somewhat of a taboo subject in the British Isles. Just prior to my appearing on a breakfast TV programme, the producer cautioned me to remember that it was morning television, a fact I needed no help recalling, as it was 5 a.m. The broadcast was beaming live from the exhibition, 'death – the human experience', at the Bristol Museum & Art Gallery.[15] Toeing the line of the accepted canonical narrative, a blurb about the exhibition said that as a society we're reluctant to talk about death and dying. The producer who was prepping us before airtime certainly seemed to believe that. Even though the broadcaster had decided to put a piece about the death exhibition on their morning schedule, she was worried that folks might be averse to contemplating their mortality whilst having their first coffee of the day. 'It's early. I mean, we know it's about death and everything, so you have to talk about it,' she said, standing in front of a giant iPhone display with a virtual memorial candle flickering on the screen. 'Just try to keep it *light*. Up tempo.'[16]

We don't like to think about death either, say the Americans. With sufficiently healthy habits, quality medical care, and better living through chemistry, maybe people can avoid dying altogether! The longevity scientists and immortalists of Silicon Valley spend their days tinkering their way towards a perfect combination of genetic, biological, and technological tweaks that will either increase the amount of time we can live healthy and productive lives, or that will make death altogether optional.[17] When the proposal for this book was being prepared for distribution to American publishers, my agent wasn't quite sure, worried that it might be an uphill battle to sell a book about death in the United States. Could I perhaps make the title a bit less ... death-y? Perhaps throw in a bit more about immortality, all the better to appeal to American fantasies of living for ever? Any psychologist worth their salt knows that avoidance is rocket-fuel for anxiety, so these conversations only led me to muse further about how it's no wonder that death taboos and death anxiety go hand in hand in the US of A. But I had a hypothesis, which was that our online existence might usher in an era when we no longer have the luxury of avoiding awareness of death.

For most of human history, the grim reaper knocked far earlier than he does today. Life was nasty, brutish and short, and death was in your face. In seventeenth-century England, the average lifespan was just under forty years; about 12 per cent of children died before they reached their first birthday.[18] Settlers in New England around the same time fared only slightly better, after some initial decimations of populations in the early days of the new colonies. As the eighteenth century waned and the Industrial Revolution cranked up successive gears, advances in agriculture and food production, and better knowledge about nutrition, made us all healthier. Cleaner water, better hygiene, and eventually vaccination made us less vulnerable to infection in the first place; antibiotics and other advances in medical science rendered many ailments minor inconveniences rather than death sentences. In only 400 years, humans have doubled their life expectancy.[19] But increasing our lifespan by 100 per cent isn't the only thing that has pushed death further away in the last few hundred years.

In agrarian societies, there wasn't much reason to stray far from where you were born unless you were fleeing plague or famine or seeking better land on which to farm your crops, tend your livestock or hunt your game. Had you wished to up sticks and move far away, there wouldn't have been a terribly efficient way of doing it. In Jane Austen's horse-and-carriage England, for example, folk were usually buried in the parishes where they'd lived, in churchyards or family plots amongst their relatives and communities. If you were a Christian, which you were statistically likely to be, you would walk past the graves of your ancestors as you left services on a Sunday, in the graveyard attached to the parish church, and everyone in town would likely remember them too. The Industrial Revolution changed all that, increasing the numbers and the mobility of people, developing ever more sophisticated transportation technologies and networks to whisk them far from their places of origin. With industrialisation came urbanisation, and in the first third or so of the nineteenth century, the population of London more than doubled.[20]

When they died, the people who'd migrated to cities might be buried where they'd ended up, not where they'd started – if there was room, that is. It didn't take long for overcrowded urban burial spaces to become inadequate for safely and hygienically handling all of the urban dead. In England's capital city, the newly forged London Cemetery Company redressed the lack by establishing large, elegant Victorian cemeteries between 1832 and 1842 – the Magnificent Seven, including lovely Highgate and fashionable Kensal Green.[21] These were stand-alone cemeteries, unlike the graveyards situated within local communities and attached to churches. A slightly later cemetery, Brookwood, even had its own rail line, the Necropolis Railway, which transported the bodies of Londoners from Waterloo to their final resting place in leafy suburban Surrey many miles distant.[22] America faced the same problem of overcrowded urban burial spaces, and Boston arrived at the same solution as London at around the same time. In 1831, Mount Auburn Cemetery was founded, becoming the first modern rural cemetery in the United States and the model for all that followed – a place where people could visit their dead in a beautiful, natural setting.[23]

The new cemeteries were born out of necessity, and no doubt they were a sensible and practical solution to the problem. They made themselves maximally attractive to live visitors by pitching themselves as 'pleasure grounds' as much as memorial spaces, complete with statuary, picnicking fields, shady trees, manicured flower beds and perhaps even ornamental, swan-filled ponds. As either an unintended or a desired consequence, however, they also served to delineate and separate the spaces for the living from the spaces for the dead. Whether your lost loved ones had moved to a faraway city or had stayed close to home, you became far less likely to walk past their resting places as you went about your daily business. Instead, you would now need to make special arrangements and efforts to visit your dead, decreasing the frequency of those visits and consigning them to the ritual rather than the everyday. Meanwhile, better understanding of infection and disease meant that the sick were increasingly cared for in hospitals, rather than at home. Influenced by the new preoccupation with public health, bodies were swiftly whisked off to mortuaries and climate-controlled funeral homes, to be handled and processed by trained professionals. Legislation was gradually introduced for what you could and could not do with the bodies of your dead.

When Sigmund Freud penned 'Mourning and Melancholia' in 1917, he promoted a linear, stepwise process of resolving our losses and saying goodbye to our dead. In many ways he was a product of his time. His stance on the subject, however distant it might have been from his own, personal experience, expressed some of the values of the Industrial Revolution: efficient functioning of society, always moving forward clear-eyed towards better, healthier and more productive lives. Life is for the living, and the dead are in the suburbs or out in the country, behind the cemetery gates. We can think about the last two hundred years as one long, gradual separation process of life and death, both in our physical spaces and in our psychology. But suddenly the worm turned – and not because anyone intended it. Over the last decade it's gradually dawning that we're in the middle of a reunion bash for the quick and the dead, and it's a surprise party that even the organisers didn't realise they were planning. It's hard to say when that party started, exactly, but it really got kicking in 2006. That's when we started putting the technology in place, in earnest, that would enable the dead to stick around in the same places and spaces that the living frequent.

Ω

TIME magazine started its 'Man of the Year' issue in 1927, changing it to 'Person of the Year' in 1999.[24] At various points in the history of the annual, it had featured collective groups of people rather than individuals: Hungarian freedom fighters in 1956, for example, and American scientists in 1960. In 2006, it surprised everyone by honouring the largest collective to date. Rendered in the grey-and-white colour scheme we now associate with Apple products, the cover image displayed a desktop monitor with a gently curving keyboard. The headline text announced that the 2006 Person of the Year was: YOU. 'Yes, you. You control the Information Age. Welcome to your world.'[25]

People rolled their eyes, complained that influential individuals had been denied their due recognition, and quickly tired of jokesters putting '*TIME* Magazine Person of the Year' on their Twitter bios.[26] In hindsight, however, it seems strange and short-sighted that anyone ever criticised *TIME*'s choice of YOU, at that particular point in history, as a superficial or gimmicky cop-out.

The web designer Darcy DiNucci had coined the term 'Web 2.0' some years prior,[27] but the depth and breadth of its impact was only truly becoming apparent in 2006. Web 2.0 isn't about any one platform or system – it's a kind of category, a description of a technology with particular affordances. It refers to those Internet platforms that emphasise user-generated content, ease of use (even by non-techie types like your gran), and interactive, collaborative capabilities.[28]

Web 2.0 levelled the playing field. With its arrival, you no longer needed to be a dyed-in-the-wool tech head or to have an MSc in computer science to easily make your voice heard around the world. In 2006, the number of weblogs – known these days as 'blogs' – had hit over 50 million, up from 23 in 1999.[29] Not 23 million, mind you: just 23. It was the year that YouTube had its first birthday, and the year that Facebook and Twitter were born, running as soon as they were foaled, like baby racehorses. Four years on, it would not be you, and each of us, feted as the most influential person of the year. Instead, in 2010, *TIME*'s chosen figure was one of the major players pulling all our strings: Mark Zuckerberg, CEO of the world's most powerful and game-changing social networking site.[30]

This new kind of Internet made it easy for even the least technologically minded of us to capture our words, images, and eventually our lives online. Once upon a time, even in developed, affluent societies, there was a far more pronounced 'digital divide' – some of us had the resources and skills, the hardware and software to be online anytime, and some of us did not. What used to be a yawning chasm, however, has narrowed into a tiny crack in the ground; in 2016, smartphone penetration in the United States exceeded 80 per cent.[31] No wonder the forces that govern our lives seem to assume everyone is connected up – pity the uphill struggle of the determined Luddite or otherwise technologically challenged individual who strives to communicate, run a business, do a bit of banking, or just stay on top of current events.

It's not just what you *do* online that matters here, however – it's who you *are* online. In this corner we have the old chestnut that one can 'hide behind' the Internet, that people misrepresent and pretend, and that you can't really trust anyone to be who they say they are. In the other corner, we have discourse two, the sometimes-satirised but increasingly accepted proposition that if someone has *no* accessible online presence, if an Internet search comes up empty, you can't really trust them either.[32] A nonexistent or ill-tended web presence can be just as problematic as a damning one. As a consequence, most of us experience the pull to store, share and utilise our information digitally and/or online. Being perceived as a proper, valid, trustworthy person in the digital age seems to come with certain expectations. A white-collar professional without an up-to-date LinkedIn profile, a plumber without any online reviews and testimonials, a musician without Twitter followers, a fashion designer who fails to produce Instagram stories, or an author whose books have no reviews on Amazon – all these people are perceived as rookies at least and suspect at worst.

So do we control the Information Age, like *TIME* magazine said, or does the Information Age control us? As you trail your tail of digital information behind you, receiving overt and subtle rewards for maintaining it and punishments for neglecting it, is the tail wagging the dog? Well, let's get a sense of where you fall on the following scale. You could classify digital-age citizens in all types of ways, but let's try arranging them on a spectrum from rejectors to enthusiasts: hermits, pragmatists, curators, always-ons, and life loggers.[33]

Hermits are the true opters-out, content in their analogue caves, and they are an increasingly endangered species. If you are holding this book in your hands, it is unlikely that you are one, but you may know one or two: the colleague who still needs to receive key information by post rather than by email; the grandparent who only has a landline and takes photographs with a film camera. They may imagine that they have no digital footprint whatsoever, not realising that information about them could be made available by other sources, even if that information is indeed basic, like work history or a list of addresses at which they've lived. Their digital footprint is more like a digital little toe, with minimal personal or sentimental weighting.

The online footprint of what I'm calling digital pragmatists tends to be deeper and more defined, but perhaps similarly impersonal. Digital pragmatists engage with the Internet and digital technologies only as much as they have to in order to get through life. They are not generally active on, or engaged with, social media in any personal way. They use connected devices for the essentials, such as banking, looking up information, working, and communicating via email.

They might or might not have a smartphone but, even if they do, they might prefer SMS – the basic text-messaging device-to-device platform offered by all mobile phones – to a specialised messenger app like WhatsApp or Facebook Messenger. While platforms such as these have caused the overall popularity of SMS to tail off since 2011, worldwide SMS text messaging is still one of the most popular means of communication,[34] and remains an option for those who do not have Internet-enabled devices. Like the vast majority of us, though, digital pragmatists are highly likely to be shooting and storing their photographs digitally. Digital-camera sales have also fallen off in recent years,[35] but that's only because of increasingly high-quality camera-phones, not because people have fallen back in love with shooting film and printing photos to put in albums. If you're a digital pragmatist, you may underestimate the size and recognisability of your digital footprint.

Curators are more cognisant of these things and so may be relatively cautious and conservative in their habits. They do everything digital pragmatists do, but also engage with Web 2.0 technologies, sharing verbal and visual material with friends, associates and sometimes strangers on social media and perhaps with the larger public via blogs or microblogging avenues like Twitter. For professional or personal reasons, they may even vlog (video blog) or do podcasts. Their digital imprint is far more personal than that of a pragmatist, and much more carefully crafted. Their selectiveness is often associated with concerns about consequences and more traditional notions of privacy. They may be digital immigrants and, as such, life in the online, Web 2.0 panopticon does not feel entirely automatic or natural to them; they may also be digital natives, perhaps people who have had negative experiences with privacy violations in the past. Context collapse[36] makes curators nervous, and that wariness influences them to consider their potential audiences and to stage-manage their online presence accordingly.

If you're not sure what context collapse means, imagine this. You're giving a cocktail party at your house and drawing up the guest list. The old friends you grew up with in the country might hate your new friends from the city, and vice versa. Your gym friends might not gel with your book group. Your partner might clash with your boss or your work colleagues. Very quickly, you conclude that there's a reason that you keep these people separate in your life, and it would be foolhardy to invite *everyone*. That's not just because not everyone would get along – if everyone were there, how would *you* act?

This is about your privacy, and not in the way you might be thinking. 'Privacy' isn't the same as secrecy and isn't just about staying apart or secluded from the gaze of others. As you move through life, you are constantly regulating your privacy, always calibrating how much and what you reveal or conceal, depending on the circumstances in which you find yourself. As you encounter each new situation, you evaluate it and make multiple decisions – some conscious, some not – about how much access to your most intimate, innermost information to grant. It's not that you are 'yourself' in one context and 'not yourself' in another. Instead, in all of your multi-facetedness, you continuously move back and forth along the spectrum between openness and closedness, in a quest for an ideal level of privacy and an optimal degree of social interaction. That, anyway, is the way that curators are used to doing it, and they carry this strategy with them from the offline to the online sphere.

But much of the online world, social media included, is a context-collapsed party, minus the cocktails but inclusive of the most motley cast of characters imaginable: everyone. The average 'always-on' is well aware of this, and it might be just the way they like it. An always-on is likely (although not guaranteed) to be a digital native, and digital natives generally approach and experience privacy quite differently. 'Generation AO' was described this way in the Imagining the Internet Center's fifth 'Future of the Internet' survey: 'By the year 2020, it is expected that youth of the "always-on generation," brought up from childhood with a continuous connection to each other and to information, will be nimble, quick-acting multitaskers who count on the Internet as their external brain and who approach problems in a different way from their elders.'[37]

The friends lists of always-ons may be more outer-circle than inner-circle and will often include people they have never met face to face. They may feel as close to online friends, people they've only ever interacted with via technology, as they do to face-to-face intimates. Always-ons are still conscious of careful presentation of their online image, but they tend to document far more of their lives than curators, and to share that documentation with more people, across more platforms. They may or may not be thinking about their digital footprint as being lasting. For those always-ons who are digital natives, they may not be thinking about this that much – young people are often less cognisant of their own mortality.

Finally, there's the life loggers, who deliberately set out to document as much of their lives as possible. Literal life logging, incorporating video capture of every minute and every interaction, was much in the news for a time, brought to us courtesy of a host of nifty wearable gadgets. Around 2013, life logging was reported to be 'all the rage. Devices like the Narrative clip, the Autographer wearable camera and of course Google Glass promised a future where we would all record every moment of our lives – for posterity or to share instantly with the world'.[38] That quote, by the way, was from a 2016 BBC news story covering the demise of Narrative (formerly Memoto), the makers of the aforementioned camera clip, with a nod to the fact that Autographer had also recently bitten the dust. Issues of privacy and consent may have contributed to the ultimate failure of those enterprises. Many people may have enjoyed using the devices, but of course it wasn't just *their* lives being logged, but the lives of anyone else they interacted with or encountered.

People attracted to life logging might have any number of driving motivations: memory problems, narcissism, paranoia, artistic motivations, scholarly research, or a wish to leave a legacy for future generations. In his novel *The Circle*,[39] Dave Eggers storied the dystopian possibilities created by life logging and life sharing, with the subsequent Netflix film, starring Tom Hanks and Emma Watson,[40] rendering those possibilities uncomfortably plausible. Reading or watching *The Circle*, you are likely to get a sense of why life logging might not have taken off (yet).

So where within this categorisation do you fall? If you're a curator, always-on or a life logger, know this. One day your physical body will be rendered invisible and silent, sequestered behind cemetery gates, enclosed in a decorative urn or scattered to the four winds, but you will likely still have a posthumous virtual self that is relatively visible, vocal and nimble. The more you participate in the online world now, whether through choice, compulsion, or even coercion, the greater the potential impact of your digital footprint after you die. Furthermore, as you've discerned by now, your digital footprint is a lot like your actual feet: extraordinarily structurally complex, with many moving parts. Break it down enough and study it sufficiently deeply, and you'll be on your way to becoming a digital podiatrist. Here are some of the main bits of anatomy that matter for your posthumous online existence: your assets, your autobiography, your unauthorised biography, your archives, and your dossier.

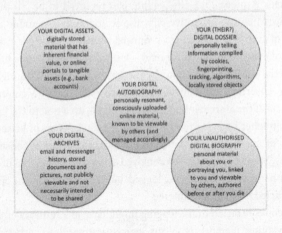

The concept of digital assets is in its infancy, and it's an albatross around the neck of nearly every estate planner, probate lawyer and will executor in the world, even those who do their level best to keep up. Traditional law makes for an uneasy fit with digital platforms, dragging them down like an unwieldy ball and chain. In the UK, the usual test for whether something is an asset that can be executed in a will is twofold. First, does it have tangibility? And second, does it have value? Most digital immigrants probably still think of assets or legacy like this, and Joan Bakewell, the Labour peeress and veteran broadcaster and journalist, was no exception when I met her on a radio programme she was hosting. 'I need to press you,' she said to the assembled panel at BBC's Broadcasting House, and then she invoked the tangibility test. 'How big an issue are digital assets? I'm an older generation, and I think of assets as furniture, books and photographs – all stuff that you can *handle*. But this is in cyberspace.' A few moments later, she upped the ante with the second test. She was on Facebook and even Twitter, but she didn't perceive the material she put on either platform as being valuable per se. 'I kind of expect it to evaporate,' she said. 'Is that part of my generation? . . . What I do know is that "assets" mean *value*.'[42]

I knew just how sceptical she was, for I'd been told by the producer that Dame Joan hadn't believed at first that the topic of digital legacy could possibly fill forty-five minutes. In only three quarters of an hour, though, it would have been virtually impossible for James Norris, founder of the Digital Legacy Association in the UK,[43] to give a comprehensive list. In service of brevity and simplicity, he divided his answer into two broad categories, the first being digital stuff of monetary value. For some of this material we might own a user license, rather than possessing it outright – our 'ownership' of the e-books, music and films that we've purchased through third-party platforms may expire when we do. Our digitally stored or published intellectual property, on the other hand – our original creative writing, musical compositions, art and photographic images – is more clearly owned and subject to the relevant laws of succession. In other words, control over it goes to your next of kin unless otherwise provided for in your will. Sounds pretty simple, right?

Delve more deeply into each of these broad categories, though, and you'll rapidly find yourself lost in a maze. To make things more complicated, the other category of digital assets that James named included 'sentimental, personal' material; certainly things that *you* might experience as valuable and tangible, but that wouldn't be considered to be assets in law. But classification is only part of the issue – another problem is the sheer volume of it all. To give you a sense of this, go online and search for a string of keywords like 'digital legacy digital assets'. Be aware, as you do this, that your search will likely be tracked and that you'll start seeing adverts for estate planning fairly shortly (more on that later). Let's look at just one of the tools that will likely pop up when you do this: the 'Digital Assets Inventory Worksheet'.[44]

This particular worksheet starts out simply enough. 'Electronic devices', reads the first category, '(e.g., smartphone, tablet, laptop computer, desktop computer, external hard drive'). Five rows are provided to list the name of the device, the user, and any password. As I do a mental inventory of the devices in my own household, and how I manage them, I rapidly realise that I not only need to double the number of rows, but I need to stock up on erasers to keep up with my ever-changing passwords. Why erasers, you might ask? Because if it were a digital document, my next of kin would need to get through a layer of passwords just to get to this document if it were digitally stored. From there the categories mount up quickly, and they don't just refer to technical *assets*, things of financial value; some are simply the digital portals to the management of those assets. Some pertain to autobiographical data – verbal, audio or visual material accessible to others, on social media or otherwise online. Some refer to your digital archive – material that isn't deliberately autobiographical and that you might never have intended to share, such as email and message histories, or documents and photos you've stored on your devices. Some things don't need to be *passed* down but merely *closed* down, like an online subscription to the *Washington Post*.

I'm willing to bet that some of the digital assets on the Digital Assets Inventory Worksheet will seem obvious to you, while others will be elements that you never imagined might constitute a major headache for someone after you've gone. Benefit accounts, like airline or other travel miles, hotel reward schemes, or loyalty programmes with different retailers. All of your email accounts, new and old, work and private. Your many financial accounts, not just your bank accounts, but credit cards, mutual funds, retirement and social security accounts, employee benefit accounts, PayPal. Online accounts with various merchants like Amazon, WalMart, Tesco, eBay. Memberships of organisations and charities. Photography and music accounts on Spotify, iTunes, Snapfish, Instagram, Flickr. Subscriptions to newspapers, magazines, blogs. All of your social media accounts: Facebook, Twitter, Pinterest, LinkedIn. Video accounts on YouTube and Vimeo. Virtual currency accounts that could have cash value, like Bitcoin. Material with value on virtual worlds like Second Life and World of Warcraft. And what about all of those documents stored on your Dropbox, Google Drive, Apple's iCloud, and other cloud storage sites? And don't forget your own websites and the domain names that you own.

I spotted a few things missing from the above list – app passwords, for example, to include hugely popular messenger applications like WhatsApp. At the prospect of filling in my passwords, I noticed a slight flinching sensation and an awareness of all the assumptions operating here. This worksheet assumes that you're happy for your next of kin or your executor to have access to *all* of your emails, *all* of your messenger communications, *all* of your personal documents. It assumes that you're content for someone else, temporarily at least, to be able to impersonate you online, even if that's for purely practical reasons. It assumes that you can fully trust that only the right people are going to be able to get hold of this document, and at the appropriate time – after you're dead, that is. It assumes that these trusted persons will be able to get into certain accounts like your Yahoo!, Facebook or iTunes before those platforms get wind that the user is dead, before their terms and conditions compel them to lock anyone else out, at least, or delete the accounts, at most. It assumes that you don't change your passwords often or, if you do, you're committed and organised enough to keep this worksheet updated.

Statistically, we're in pretty unlikely territory there. According to many estimates, fewer than half of adults have made a will.[45]

How many do you imagine have made arrangements for the administration and disposition of the digital portions of their estate? But here's one thing that you can safely assume, in the exceptionally fast-moving world we live in: the foregoing list of potentially problematic assumptions, like the list of digital assets that preceded it, is *not* comprehensive.

Adam shared something in common with the overwhelming majority of us – he never completed an inventory of his digital assets or a list of his passwords.[46] One day, Adam suffered a brain haemorrhage and fell down the stairs, a series of events that proved fatal. Because the door was unlocked and the circumstances unclear, the police treated it as a potential crime scene, and they assumed control of his phone. His daughter, Kim, had just arrived on holiday when she got a call from her friend Martin. Martin had never met Kim's dad, but had discerned something was wrong through a complex relay of information involving social media. 'I don't like Facebook a lot, I don't think it helps people,' said Kim, 'but actually it was the only way that we found out.' Thanks to Facebook's mutual-friends feature, Kim was able to reach her father's side before he died.

Kim had probate and assumed responsibility for managing Adam's estate, and at first it seemed straightforward. All the traditional things were easy – taxes, banks, gas and electricity accounts. Then it came to the digital side of things. James Norris of the Digital Legacy Association reports having heard multiple anecdotes about people seeking the cooperation of morgues to unlock their loved one's fingerprint-enabled smartphones, although when I discussed this with Dr John Troyer of the Centre for Death and Society (CDAS) at the University of Bath, he figured that this was highly unlikely to be successful.[47] Phones that use fingerprint touch technology are unlocked using capacitive touch, reliant on the electrical charge generated by the skin of a live person. In any case, there were no such issues with access here – Adam's phone didn't even have a passcode. 'I could hack all of his accounts,' Kim said. 'I sat there with his iPhone, and I hacked his Apple account, I hacked his Amazon, I hacked everything, and accessed it. The only reason I could hack all those was because I could get into Gmail – I could reset all the passwords through that.' Then it got more complex. Adam had been a bookseller, and he had plied his trade exclusively online, on platforms like eBay, Amazon and a handful of dedicated book sites.

'When he died,' Kim said, 'there were still orders coming through. We searched for the books and managed to fulfil the orders, because it seemed the right thing to do and it was easier than trying to cancel them.' But Kim didn't want to continue Adam's business; she wanted to close it. So she took down the unsold listings, leaving an empty shop front. If she hadn't done this, orders would have continued coming in and could have gone unfulfilled, possibly leading to poor reviews. 'That would have blighted his memory a little, if something unjust were written and it was out of his control,' Kim said. So she was happy that her Gmail-facilitated hacking enabled her to log in and empty Adam's online inventory. But that wasn't enough. 'I wanted his store fronts to be closed down. I don't know. I just wanted that to be so. Does that make sense? There was no need for them to be there any more.' To do that, she needed to get in touch with the relevant platforms. And she did. Again, and again, and again.

'I would ring. They would tell me one thing. I would ring. They would tell me another thing . . . . And it just took weeks and weeks and weeks and weeks. I hacked into LinkedIn and switched it off, but it's not switched off, and they still send him reminders. I rang. They said, oh, but it's closed down. But it's not. Same with Amazon and eBay. There was no way they would have even spoken to me if I hadn't had all the details and access to emails. I couldn't have even accessed the seller helpline. Unless you are that seller, you can't access the seller helpline. It seems very hard to actually remove people's digital trace.' Kim rummaged for her phone and tapped in a few things. Turning the screen towards me, she showed me Adam's empty shop fronts. 'See?' she said. 'They're still there.'

'Wait,' I said. 'How long have you been trying to sort this? When did he die?'

'One and a half years ago,' she said.

Ω

Unlike Adam's online business and other digital assets, your digital autobiography may have no monetary value whatsoever. There may be exceptions; who knows, representatives of your estate could compile the posts on your blog and turn them into a smash-hit bestseller. Generally speaking, though, you can think about your digital autobiography as whatever you deliberately 'publish' online, to an audience, about you and your life. Most social media and online publishing forums now accommodate multiple types of posts: verbal and visual; text, audio and video; original material or links to others' material. Whether you are spare and selective in what you post, more of a curator, or whether you provide a continuous stream of updates like a life logger or a particularly active always-on, you largely have creative control over how you shape and edit your digital autobiography. It represents you as you wish to appear to others in the world. Being co-constructed with other people through conversations and 'likes', it also shows you as we all are in life – embedded in the context of our relationships. Traditional obituaries and eulogies, written and delivered by someone else, could easily portray you as better, worse or different than you were without your being able to do much about it, but a posthumously persistent online self-portrait makes it considerably more difficult for others to seize control of how you are remembered. The online dead retain an ongoing measure of control over their lasting image, and that image may be vivid and rich indeed – negligible in monetary value but nevertheless priceless to many, particularly when hardly anyone prints out photographs or commits words to actual paper any more.

Nick Gazzard lost his daughter Hollie in 2014, under horrific circumstances that will be explored at length later in this book. He told me that his daughter had something like 700 friends on Facebook, which 'back in those days was quite a lot of friends'. When Hollie died, he was a social media novice himself, so had hardly ever seen her profile. After she was gone, though, he did explore it, and was tremendously moved by what he found. 'It was her legacy,' he said. 'That was her legacy. If I want to know more about Hollie, I'll go onto her Facebook site. I didn't know that before, because I wasn't a

Facebook user, but when all this came about, I set up an account and had a look at it . . . everything's on there. That's Hollie's life. It felt like her telling me, these are the things I've been doing. Things I didn't even know she did, the things she said to people, some of the laughs that she . . . well, it gave me a different perspective on what Hollie was like. I could see . . . the joy. The joy of Hollie.'

The fact that sites like this have an autobiographical function is not lost on companies like Facebook. It allowed memorialisation of profiles – in one form or another – as early as 2007, pledging that they would remain as sites of mourning and remembrance. While it was once built to focus attention only on recent events, it broadened the horizon in 2011 and launched the new Timeline design, explicitly positioning itself as an autobiographical instrument, a place to tell and store one's whole life story, right back to birth.[48] A few years later, in 2015, it introduced a feature allowing people to nominate someone to manage their profile page after death, and the name they chose for that feature was, again, significant: Legacy Contact.[49]

For many or most of us, using social media is about living our lives, rather than anticipating death and considering legacy. For those of you who *are* aware of the legacy element, however, unless you are keeping constantly abreast of evolutions in terms and conditions, and planning your digital estate accordingly, you may find that you've made some incorrect assumptions about the longevity and integrity of your digital autobiography. If your Facebook profile contains the whole of your history and every important photograph you've taken since you were sixteen, and you feel 100 per cent certain that your Facebook profile will remain intact ad infinitum for friends, relatives and descendants to remember you by – well, let's just say that some of the stories in the rest of this book may send you running to treble-check the site's terms and conditions in general, and your own account settings in particular.

Ω

While it's lovely to have the illusion that you're completely in control of your personal information and public image, the biographical material that you select and upload yourself is probably not the only stuff about you that's floating around on the Internet. Have you Googled yourself lately? Maybe you should, because you'll likely spot information about you stemming from all manner of sources. Unless you're a celebrity, with writers vying to publish the definitive story of your life, the unauthorised biography you see represented in the search results will be a fragmented *bricolage* rather than a coherent narrative. I'm guessing that you'll feel OK about some bits of it and may even be alarmed about other bits. You are likely to find information that is inaccurate, or deceptive through its being out of context, or entirely accurate but not something you want published, or just unexpected. You could be surprised because you've forgotten a few things, or because you were unaware that the information was captured in the first place. There might be information that appears to be about you but is actually about someone else. You might even discover that you're dead already, that you've been dead for four years, which is what happened to Rachel Abrams.

Whether it was algorithms or humans (or both) that were responsible for the initial mistake, and the outward ripples from that original error, proving that she was alive turned out to be rather vexing for Rachel. In the *New York Times* in December 2017, she described her efforts to convince Google that, as Mark Twain famously said, reports of her death had been an exaggeration.[50] 'Plenty of people try to remove negative or inaccurate information about themselves from the Internet,' she wrote. 'There are entire companies that will do this for you. But often, the misinformation appears on websites other than Google, which Google doesn't really see as its problem.'[51]

Another element of unauthorised biography is the material that people write about you after you are gone, which also forms part of your digital legacy. People might share their fond memories of you on your posthumously persistent social media sites. They might write lovely obituaries on an online memorial site like World Wide Cemetery,[52] the Internet's longest-running such place, or pen their remembrances in the funeral home's online guest book. They might create and upload a memorial video on to sites like YouTube or

Vimeo, or create a whole website just to remember your life. This is all rather nice, but unfortunately sometimes it's rather nasty, as you'll read later on.

Ω

The phrase 'digital archive' could refer to all sorts of collections, but for the purposes of this book and this classification, digital archives are different from the autobiographical and biographical information that's published online to an audience. Instead, digital archives are the behind-the-scenes material: our emails, our SMS threads, our messenger-app conversations, and the documents and images we store on personal devices or in password-protected cloud servers. While we might share these with others (particularly, of course, the people with whom we're conversing), they are private rather than public data. Sometimes they are pedestrian and administrative, sometimes impersonal and sometimes exquisitely revealing, but whatever their nature they were likely never intended for wider dissemination and may even be at dramatic odds with an individual's preferred public persona. Whether you're an open book, or whether you'd sooner cut off your finger than give your significant other access to your file of passwords and access codes, I suspect most of you would hesitate slightly if you thought that *everything* in your archive could be seen by your next of kin after death. But that couldn't happen. Could it?

Sometime around the start of the new millennium, a teenager from Michigan named Justin Ellsworth opened an email account with Yahoo!. When setting it up, he would have gone through a routine that is drearily familiar to most of us: he would have had to tick a box or two, indicating his acceptance of the terms of service. Three quarters of us admit that we don't read terms and conditions at all,[53] but if Justin did happen to be among the conscientious 27 per cent, he would have read that the account was non-transferable. Upon his death, his Yahoo! ID and the contents of his email would not be handed over to anyone but would be deleted once Yahoo! determined that he was deceased. Maybe he didn't worry about this. He was young and fit, and young and fit people tend to believe that they'll live for ever. In any case, though, the clause was legally binding and clearly worded, and Justin would have had every reason to believe that Yahoo! would do what it said it would in the event of his death. That's not what happened.

Justin Ellsworth didn't live for very much longer. He became a Marine and went to fight in Iraq, where his life ended with a roadside bomb in Fallujah in 2004. In addition to his service and sacrifice for his country, what he is remembered for now is a court case that became the first major test of post-mortem privacy in the digital age. Justin was unmarried, childless and intestate, and the personal representative of his estate was his father, John. Justin had never shared his Yahoo! password with his dad, but John was adamant that he needed access to the account, for reasons that were unclear. While the media reported that it was for sentimental reasons, to make a scrapbook of his son's time in Iraq, elsewhere John said that important financial information was stored in the account and was needed for management and disposition of Justin's estate. Yahoo! retorted that since Justin had signed up to their terms of service, if John Ellsworth wanted access to Justin's account, he'd have to take it to court. So he did.[54]

Part of the reason that accounts are non-transferable and login details are protected is to prevent impersonation. Service providers usually contract with individuals, and a group of people cannot generally share rights and entitlements to a single email account. There was also the little matter of the federal Electronic Communications Privacy Act,[55] which underpinned the non-transferable, destroy-upon-death stipulations of Yahoo!'s T&Cs. Unsurprisingly, therefore, the courts did not initially grant John Ellsworth access to his son's account.

In light of all that, what happened next was rather surprising. The family appealed, and the court relented. They still refused the estate direct access, but they ordered something that was tantamount to that, at least where Justin's privacy was concerned. They ordered Yahoo! to download the contents of Justin's password-protected account – photographs, documents and emails – into an accessible format and to deliver these to his father. Shortly thereafter, John received three large bankers' boxes and a CD-ROM containing approximately 10,000 pages of material.[56] Certainly, much of this data was not suitable for scrapbooking Justin's sojourn in Iraq, and nor was it relevant to the finances of his estate. Indeed, many of the emails were to or from people John Ellsworth had never heard of, but whose correspondence with his son he was now free to read. Like the material that is actually visible about us online, the content of our email accounts is context-collapsed as well.

The judge's decision in the Ellsworth case remains controversial, and it is unlikely that we will ever know for certain the key arguments that persuaded the judge to order Yahoo! to hand over the emails to Justin's father. Rebecca Cummings, an Atlanta-based expert in wills, trusts and estates, and a former law professor, felt sufficiently strongly about the Ellsworth case to write a paper arguing that this should never have been the outcome.[57] While many justifications for the access may have been cited, she wrote, none of them were 'compelling enough to override [Justin's] intent in a probate matter, or to grant personal representatives[58] default access to ... password-protected email'.[59] Justin Ellsworth, if he were looking on from the great beyond, might not have been too pleased about it. And what about you? How readily would you fill in a worksheet full of passwords to grant your executor, next of kin, or any other person full and unfettered access to your entire digital archive?

Ω

So far, I've talked about the digital footprint that we know about, more or less, even if we don't consciously think about its becoming part of our digital legacy. The remaining category – the digital dossier – is a different matter altogether. Although we're increasingly aware of online surveillance technologies, we may be unaware of the sheer extent to which we are being silently watched, and tracked, as we travel about the Internet. The digital fragments that we trail behind us in the digital world are shed as inevitably and unintentionally as the invisible cascade of skin cells we leave in our wake offline. If you think that cookies, algorithms, fingerprinting and tracking technologies cannot really give an accurate picture of who you are, you and I clearly aren't having the same experiences.

One day, I was thinking about how I needed some new boots. I was envisioning some versatile leather boots, in black, probably about knee height. Despite not having mentioned this to anyone or conducted any Internet searches, I was suddenly bombarded with adverts for exactly the sort of knee-high boots I'd been envisioning. This has gone too far, I complained on Facebook, describing these events. The consequence of this post, unsurprisingly, was even more advertisements for boots. Perhaps my longtime use of the search engine Google had made my needs exceptionally easy to predict. '[T]he search giant is very good at tracking you and using algorithms to anticipate your needs,' says an article in *Lifehacker*.[60] Clearly.

And it isn't just Google that's spying on you. Every time you click 'like' on Facebook, check in at a location, or use a recommended link on social media to go and purchase something, you are giving legion third parties vital information about yourself. If certain settings are enabled, or default settings not disabled, your smartphone or smart watch is invisibly logging and even disseminating all manner of personal data, without your ever being aware. Even if you don't check in, many of the apps you have on your phone can use your photographs to identify your location.[61] A few years back, British technology journalist Geoff White and security researcher Glenn Wilkinson proved these points by alarming audiences around Britain with an interactive stage show entitled The Secret Life of Your Mobile Phone.[62] If you watch video clips of their demonstrations, readily available online, observe the facial expression of one hapless volunteer when Geoff and Glenn were able, without ever touching her mobile phone, to tell her what cafe she'd visited recently in Amsterdam.

Maybe you think that this sort of information could never be meaningful to anyone left behind, but let's consider just one thing that is routinely tracked, for your convenience and your more efficient browsing experience, of course: your search history. The handful of keywords that you enter when looking for information online.

Really? Just how personal could this really prove? Well, in 2006, the un-redacted search data for over 650,000 AOL users were released to the public at large, rather than just to the academic researchers for whom they were intended.[63] While a small collection of your searches might not prove particularly revealing, taken in aggregate over a period of time, a string of searches is capable of exposing the inner workings of the mind, the struggles of the heart, the dark night of the soul. These are unintentional, psychologically incisive autobiographies, written by people who don't know they are authoring them.

You will find no better illustration of this assertion than *I Love Alaska*, a 2009 documentary film by artists Lernert Engelberts and Sander Plug.[64] *I Love Alaska* consisted solely of one user's search data from the AOL leak, read out against a backdrop of isolated northern landscapes. In a southern American accent, a female voice narrates three months' worth of Searcher #711391's searches, unfurling a heartbreakingly poignant audio novel in the process. The blurb for the film, divided into thirteen minimovies, describes it as a story that 'reveal[ed] the yearnings of a conflicted housewife who dreams of escaping a life of obesity, sexual frustration and skin trouble in the searing Texan heat by running away to Alaska'.[65]

Although she was never identified by name, we know that she was a middle-aged, married, surgically menopausal woman from Houston, with weight issues and questions about her sexuality. Searcher #711391's style and phrasing when engaging with search engines helped boost her history into epic-poem territory rather than just being a list of boring words, and the sequencing of searches was as telling as their content. Friday 21 April 2006: 'Back pain' . . . 'How to make good impression first time meeting an online friend' . . . 'Breast orgasms'. Saturday 22 April 2006: 'What is a womaniser' . . . 'Can you use prepaid phone cards on cellphones'. Sunday 23 April 2006: 'How to get rid of nervousness of meeting a blind date' . . . 'How can you tell if someone online is lying'. Monday 24 April 2006: 'Symptoms of heart attack' . . . 'Trapped gas in chest' . . . 'Never admit to an extramarital affair'.[66] Imagine if it were Searcher #711391's husband, rather than two filmmakers, perusing her search history after she died. What might he realise that he hadn't known before?

Even a pithier, more generic searching style can still prove personally telling, though. Searcher #4417749's history betrayed less emotional pain and suffering, and her style was less verbose – 'landscapers in Lilburn, Ga' and 'numb fingers', for example – but that didn't prevent the New York Times from swiftly identifying her as 62-year-old widow Thelma Arnold. 'My goodness,' Mrs Arnold said to the Times reporter who showed her the search history. 'It's my whole personal life. I had no idea somebody was looking over my shoulder.'[67]

The digital dossier is the element of digital legacy that few people think about ahead of time; the digital traces that go beyond the more obvious, showy Facebook profiles, the blogs, the digitally stored photographs. Never intended for presentation to others, such material may be all the more revealing of the people behind the social masks that we all assume. When Vered Shavit lost her brother and went through his laptop, she found it emotional to witness the very way he had organised his desktop, the names he had given to his files – it was a window into how he reasoned, thought, prioritised, a portal into his cognitions.[68] When Kate Brannen lost her mother, she often returned to her mother's computer whenever she wanted to feel close to her again. Once again, these shards of unintentionally autobiographical material were amongst the most moving things on the laptop. '[H]er computer activity was like a breadcrumb trail through her inner life,' Kate wrote, 'her interests, her hopes and her plans for the future, even those that would never come true. The bookmarks in her Safari browser served as a compass on a journey into my mother's mind.'[69]

Ω

So unless you're one of the nearly extinct digital hermits, you have a digital footprint. Unless you go to some considerable trouble to prevent it, that footprint will one day form your digital legacy, for good or for ill. For each day you log into email, for every time you're on your smartphone, for every hour you fritter some time on the

Internet and for every status post you put out there on social media, you're penning another line of your autobiography. You're laying another brick in the edifice of your future mausoleum. And you're not building it in some separate digital cemetery, you're constructing it smack dab in the middle of the bustling metropolis that is the Internet of Web 2.0. The dead have returned from the pleasure grounds in the suburbs, mingling in the community again. They're everywhere you go online, their images on your screens, their voices in your ears – accessible anytime, in the palms of our hands. Does that change what mourning and grieving is like in the digital age? I have two answers for you. The first answer is no, not at all. And the second answer is yes, you'd better believe it does.

Chapter Two
The Anatomy of Online Grief

If you're taking a plane to Ireland from an English airport, you'll find that you start your descent at right about the moment you reach cruising altitude. Arriving at your destination, you will hardly experience a revelation to the senses – architecturally, topographically and meteorologically, Ireland feels a lot like many parts of England. Less visible to the naked eye are the cultural differences, the often-radical step-changes in outlook that are so easy to underestimate when there's a common language. Less than two hours' travelling time can catapult you into an entirely different culture of grief.

The first time I crossed the Irish Sea, I was going to speak about mourning on social media for some organisations in Dublin.[70] It was 2014, a few years after my encounter with the Glasgow Spiritualist, and online mourning was now a reasonably well-known phenomenon. I was a bit taken aback, therefore, when it garnered a considerable amount of media attention, and I voiced my surprise to one of my hosts as she was driving me to do a talk at the Psychological Society of Ireland. I think this feels quite new here, she said. Maybe we're just a bit behind? Hmmm, I thought. I wasn't sure why that would be. One of my PowerPoint slides that weekend featured a graph plotting Facebook usage in various countries, and there certainly wasn't that much of a difference between Ireland and England.

But then again, wherever I go, especially when I'm speaking with digital-immigrant audiences, I still encounter some level of concern about going online to grieve. Radio and TV presenters express their reservations in exaggerated, emotive ways, turning up the provocation to keep the audience tuned in: isn't this all a bit *creepy*? I mean, it's pretty *morbid*, talking to dead people online, isn't it? Everyone asks questions that pull for binary, black-and-white answers: is it good or bad to mourn online? Is it healthy or unhealthy? Should we be worried or not?

When I arrived at the psychological society, I fully expected to hear some of the same kinds of questions. Psychologists and other practitioners well versed in grief and mourning tend to simply frame their scepticism in a more sophisticated way, such as whether interacting with digital remains in mourning carries a risk of 'complicated grief'. Sure enough, this group of psychologists raised the usual issues, but there was something additional repeatedly rearing its head. It struck me because I hadn't heard it expressed to the same extent anywhere else. I was in the midst of pointing out a particular benefit of grieving online, which is that the bereaved can have access to a community of mourners twenty-four hours a day, seven days a week. A young woman – surely a digital native herself – raised her hand suddenly and with the air of having a very definite point to make. 'I don't see how that can be helpful,' she said. 'That can't be good. They should be with their *families*.'[71] Others nodded, chiming in with similar comments. Yes, surely people in grief should not be on their devices, in their rooms, interacting with digital remains and talking to dead friends and fellow mourners online. It was inappropriate and possibly downright wrong. They should shut their laptops, put down their phones and *go and grieve with their families*.

I thought that I understood what I was hearing here. I was familiar with the attitude, particularly amongst psychologists and other kinds of mental-health practitioners who privilege face-to-face interaction. I'd even done a study of English psychologists' attitudes to digital technologies, and one of the major findings was that, whatever their age, they saw online interaction as an inferior type of relating, somehow not 'real', a substitute for authentic contact.[72] In possession of these recently gathered data, I was confident in my response and encouraged people to reflect upon the underlying, knee-jerk beliefs they had about the 'right' thing to do in the face of grief.

'You're making several assumptions here,' I said. 'You're assuming that family members will be *able* to speak about the death, when they may be struggling with grief themselves. You're also assuming that they'll be *willing* to speak about it – people often believe that children and young people should be protected from death. The online environment might be the only place where someone feels comfortable expressing their grief or talking about the dead person; there may be spoken or unspoken rules against that at home.'

In other words, while I was exhorting them to reflect upon their assumptions, I was blissfully unaware of how many I was expressing myself. Climbing back into my host's car after the talk and grappling with a sense of *déjà vu*, I told her that I might have misjudged something about my audience. 'Can you tell me a little bit about death and grieving in Ireland?' I asked her. It was a topic that I hadn't researched before my arrival, because I didn't realise that it would be particularly at odds with what happens in England or the USA. I don't remember her precise words, but what she described was indeed utterly different from anything I knew. I had been describing online mourning as an antidote to the silent, isolated suffering that can come with death, but my Irish audience had reason to be flummoxed. They were trying to understand why a cure was needed for an ill they didn't recognise.

In 2017, Kevin Toolis published a book called *My Father's Wake: How the Irish Teach Us to Live, Love and Die*.[73] In an article for a London newspaper drawn from that book, he writes, 'In the Anglo-Saxon world, death is a whisper. Instinctively we feel we should dim the lights, lower our voices and draw the screens. We want to give the dead, dying and the grieving room. We say we do so because we don't want to intrude.'[74] In contrast, Toolis's own father, Sonny, had a traditional Irish wake. The body was cared for and kept at home, with the children playing in the sitting room at the foot of the coffin. People came from far and wide to socialise and gossip, eat and drink, and commiserate with the family in the dead man's own front room, alongside his body. Wakes like this used to be a more widespread phenomenon, but Toolis identifies the same forces discussed in the last chapter – urbanisation, medicalisation, industrialisation – as reasons for their steady decline in much of the West. But for whatever reasons, on Irish soil, the Celtic traditions persist; an 'ancient form of death sharing',[75] as Toolis describes it.

My Ireland visit reminded me how hard it is to wiggle free of the automatic assumptions about grief, which may be highly local, particular to our culture, our religion, our community, or our own family. We assume that we know what constitutes 'healthy' grieving. We have opinions about what an appropriate legacy looks like, what a proper send-off is, how it's acceptable to respond to a bereaved person, or how we should behave in the face of death. We have our own experiences of what has and hasn't been helpful for us. So, with our personal looms already strung with the monochrome warp of our beliefs about death, we're now holding in our hands the multi-coloured, jumbled weft of a much newer material: digital technologies. Our assumptions about the latter are highly variable, connected to too many contextual factors to count. Life online is either destroying our communities or fostering our connection with them. Constant connectivity either promotes isolation or prevents it. Social media either inspire us to be prosocial or turn us into rampant narcissists. And digital remains are either good for us in grief or not. What is the anatomy of grief in *digital* culture, within *digital* communities, and what rules are we meant to follow?

In February 2013, there was a story in the Australian press about the digital legacy of teenager Allem Halkic.[76] After being cyberbullied via SMS messaging and on MySpace, Allem threw himself off a bridge in 2009. The ensuing trial was the first cyberbullying case to come before the Australian courts.[77] After he died, his father Ali used to go into his son's closet, inhaling the scent lingering on his clothes, but after a while the smell faded – it became 'dust', as Ali put it. But his parents then discovered something that wouldn't fade: Allem was still logged onto his Facebook account. Ironically, this digital artefact became one of the most precious things to his parents, helping them connect with a son driven beyond the brink by the cyberbullying he'd endured. 'There is no way in a million years we would disconnect his Facebook profile, just no way,' Ali said in the news story. 'In the first year or so after he died, he was getting a message a day from friends and then it became slower. I got disappointed when people didn't write because I didn't want him to be forgotten.'[78]

The author of the article remarked that no one could dispute the value of persistent social media accounts to bereaved family and friends,[79] but he was wrong about this. Just a couple of months after this news story, the BBC reported that a grieving mother in Brazil had taken Facebook to court over the memorial page for her daughter. The precise things that had brought Ali and Dina Halkic joy were too difficult for Juliana Campos's mother to endure. 'This "wailing wall" just makes me suffer too much,' Dolores Pereira Coutinho said. 'On Christmas Eve many of her 200 friends posted pictures they had taken with her and recalled their memories. She was very charismatic, very popular. I cried for days.'[80] Facebook tried the middle way first, making the tribute page friends-only, but that wasn't enough for Mrs Coutinho. The thing most feared by Allem's parents and most craved by Juliana's mother was the same event – the permanent disappearance of their child's Facebook profile.

Unsurprisingly, research indicates that people who regularly use social media find it to be more useful in grieving, while those who don't are more likely to harbour unease and doubts about whether digital remains should continue to exist:[81] a 'coping paradox'.[82] The same digital artefact, such as a posthumously persistent Facebook profile, could mean inestimable comfort for an always-on, or wracking emotional pain for a digital pragmatist. How you experience digital remains, and whether it will be helpful for you to access them in your grief, will be determined by your relationship with digital technologies, woven together with your experience of a particular bereavement. If the last chapter got you musing about the former, the next bit of this chapter is designed to help you think about the latter. Before we go there, though, here's a caveat that will, by now, come as no surprise to you: in matters of grief, little is predictable, and nothing is definitive. Ask whether interacting with digital remains is a 'good' or 'bad' thing, and don't expect a yes-or-no answer.

Ω

Many of us worry that we're not feeling or behaving correctly in the face of a loss, or perhaps we get concerned about other people, for the same reason. Open any search engine and enter the word 'grief'. If you're using Google, notice what it says in the 'People also ask . . . ' box, for this is a window into common preconceptions about mourning. In the popular searches, you can discern the things about grief that people take for granted, but about which they can't recall the precise details. On the day that I search, first on this list is, 'What are the seven stages of grief after a death?', closely followed by 'What are the stages of grief?' and 'What are the five stages of grief and loss?' Looking at the search results themselves, you won't have to scroll very far down, if at all, to encounter multiple references to stage models of mourning.

The idea that *the* grief process consists of a predictable sequence of stages, eventually ending in acceptance, has brought comfort and reassurance to untold numbers of bereaved persons since 1969, when Swiss psychiatrist Elisabeth Kübler-Ross proposed it in her seminal book *On Death and Dying*.[83] Kübler-Ross's five stages – denial, anger, bargaining, depression, and acceptance – weren't actually even based on bereaved people. Instead, they referred to the stages she had observed dying people *themselves* go through. Nevertheless, they were rapidly applied to the grief of those left behind. The reading public seized upon her stages with a kind of hunger, as though they had been waiting for half a century for someone to explain how to do the 'grief work' that Freud had talked about in 'Mourning and Melancholia' in 1917.[84]

If people have been confronting loss and grief since the dawn of time, why the desperate uncertainty, all this searching for a proper course of action? Perhaps getting grief right has seemed particularly important since the mid twentieth century, when psychiatrist Erich Lindemann took the vague explanations of grief work in 'Mourning and Melancholia' and translated them into a modern psychiatric diagnosis.[85] Thanks to Lindemann and the burgeoning medicalisation of human experience that has only grown steadily since his own efforts, failing to cleanly sever ties to your dead within a seemly period of time could land you in the doctor's office. You might even leave carrying a prescription for antidepressants and be labelled as suffering a 'major depressive disorder'. Once upon a time, if someone had recently suffered a loss, this would have kept them from receiving a depressive-disorder diagnosis. However, in the fifth incarnation of the *Diagnostic and Statistical Manual of Mental Disorders*,[86] published in 2013, major depression no longer features a 'bereavement exclusion'.[87]

Quite apart from the concern about going crazy, or about being perceived as going crazy, you can understand why we might cling to the idea of an orderly succession of grief stages, unfolding one after another until you reach the point when you let go, or when grief lets *you* go, once you've served your term, put in the work, done your time. Human beings struggle mightily with those things we cannot predict or control – it's not coincidental, then, that unpredictability and uncontrollability are two of the main ingredients in the mix when we experience stress and trauma. Imagine a bereaved person in the throes of unimaginable emotional pain, on their smartphones at 2 a.m., wondering how long they will have to endure this, trying to find out what they can expect to happen next, straining to see if there is any light at the end of the tunnel. Maybe that person has been you, at some point. You wouldn't even have to tap out 'grief', because autocomplete provides that as suggestion number one, as soon as you enter 'stages of'.

The fantasy of predictability, and the hope and sense of control that it brings, helps explain why Kübler-Ross's stages continue to dominate the discourse, on the Internet and off, even though they've long been regarded as a comforting fiction, 'practically folklore'.[88] Part of what's implausible about the stages is the idea that they occur chronologically; later in her career, Kübler-Ross tried to loosen people's expectations around that by claiming that she'd never intended the model to be interpreted as linear. But the popular imagination is drawn to simplicity, and perhaps the thought of sliding back into that pain again, having thought you had reached the end, is too disheartening for many to contemplate. In our darkest hour, we hope that what Freud reassured us about is true: grief is supposed to come to an end, and it will, if you're doing it right.

Despite the strength of these canonical narratives of how we 'do' grief in the West, more and more people came to realise that perhaps we really had lost our way with death. By the mid-1990s, there was a groundswell of voices arguing for that idea, collected together in an anthology about a concept called 'continuing bonds'[89], mentioned briefly in Chapter One. Continuing bonds theory was the most not-new new thing you can imagine, given that continuing bonds with the dead has existed throughout history and even pre-history, across most cultures.[90] Only with Freud had the West started to properly veer off on the slip road. Continuing bonds was a shot across the bow to the 'curious notion'[91] of grief work – a construct that argued persuasively that continuing relationships with the dead wasn't worrying, it was just part of being human. Westerners who thought it was critical to move on and obligatory to sever ties were being told something that people in many other cultures have maintained throughout: keeping close to your ancestors is normal, adaptive and usually positive.

Continuing bonds are a bit of a free-for-all, in contrast to the buttoned-up positivism of medical and stage models of grief, and they happily morph to fit the context of each culture, subculture, family and individual. We may experience continuing bonds purely internally, in the form of memories or feelings of connection to the dead that aren't witnessed by anybody else. They may play out in our behaviour, whether our actions occur privately or as part of highly public rituals. In China, continuing bonds with particular ancestors may be non-negotiable, for funeral and remembrance rites are needed to ensure that your dead are content, happy ancestors instead of hungry, angry ghosts. If you're from a more individualistic culture, continuing bonds may be more organic, personalised and spontaneous, and you might only do it with those to whom you feel most closely connected. That could mean family, but not necessarily – you might retain a strong sense of connection to people to whom you have no blood ties whatsoever, whether those are friends, or people you've never met, like celebrities you admire or other role models who inspired you. You might think of a dead person as a guardian angel who looks after you, or as a soul that still touches your own.

Obviously, cultural context matters a lot here. The sociologist Tony Walter explains that some us have a 'care culture' of mourning. In care cultures, people believe that the dead still exist in some kind of spirit realm, and that they continue to need us to look after them in some way. He contrasts this with 'memory cultures', which hold that we must accept that the dead are gone, but in which we're encouraged to remember and honour their legacy, and to think about them as living on through their descendants. Memory cultures don't generally see the dead as separate entities with their own ongoing agency. East Asia and the global South are care cultures, and Western European cultures tend to be memory cultures. 'It is hard to talk about the European dead,' says Walter, 'without using the language of memory.'[92]

Whether you come from a care culture or a memory culture, however, it all sits under the umbrella of continuing bonds. If you've ever lost someone close to you, you've probably experienced an ongoing connection with them yourself, in some form. You hear their voice, catch sight of them, feel their physical presence. You feel their guidance or support in your thoughts, your dreams, or the events of your life. You carry them with you – not just their memories, but their values, their characteristics, even aspects of their personality. You might make conscious efforts to do things as *they* would have done, or go places they would have visited. Emotionally and psychologically, they remain part of who you are, and they still play a part in the systems around you. 'Living people play roles, often complex, within the family and psychic system', write the editors of the new edition of *Continuing Bonds*. 'After they die, roles change, but the dead can still be significant members of families and communities.'[93]

That quote explains why you can *expect* every single bereavement to be necessarily, inevitably unique. If you have four siblings, you don't have a generic 'sibling' relationship with each of them. If you have 200 Facebook friends, you don't possess an equivalent bond with every one of them. The relationships you share with your fellow humans may have similar themes and characteristics, given that the common denominator is you, but every relationship varies. Because relationships are unique in life and just as distinct in death, every loss experience will be different too, even if it can be classified within a particular type, or trajectory. Grief scholar George Bonanno, for example, has charted three patterns of grief reactions in bereaved people. Those with chronic grief reactions are so overwhelmed by loss that they may struggle to function in life for years; people who gradually recover are hit hard but eventually manage to reassemble and carry on with their lives; and the resilient are 'shocked, even wounded, by a loss, but . . . manage to regain [their] equilibrium and move on'.[94] This doesn't mean, however, that each of us is innately a particular 'style' of griever, for depending on the loss, you may experience resilience, recovery or chronic grief.

Within each of these trajectories, however, there's a common characteristic – the intensity of grief judders up and down. Grief has peaks and troughs, waves of sorrow alternating with periods of contentment and even joy. Not only is it normal for these oscillations to happen, but it's typical for them to be utterly unpredictable and uneven. Some theories of grief attempt to provide a framework for this experience; the dual process model, for example, describes how we swing back and forth between loss orientation, when we're preoccupied with the dead person, and restoration orientation, when we're focused on aspects of life that don't include them.[95] But Bonanno thinks that even this is too rigid, not giving sufficient credit to the sheer amount of oscillation that occurs. '[W]hen we look more closely at the emotional experiences of bereaved people over time,' he says, 'the level of fluctuation is nothing short of spectacular.'[96]

The recap of all of this is pretty simple. Even though you may personally resonate with Kübler-Ross's stages, there's no right or wrong way to grieve. Whatever your process is, there is nothing abnormal about an unpredictable panoply of moods, or an oscillation between loss and restoration, between being sunk in grief and carrying on as though it never happened. Continuing bonds with your dead, in all its myriad forms, is par for the course, just as it has been for millennia. Whether you are actively caring for and interacting with your dead, or simply remembering and honouring them, it's still continuing a bond. Everyone's needs and desires are different, and that's OK. The choice of how you continue bonds, or not, should be up to you, not dictated by anyone else. That brings us to the sticky bit – the digital age has made continuing bonds easier and harder all at the same time.

Ω

Let's start with the easy part: the digital environment is perfectly designed to facilitate continuing bonds. But is it a medium that supports a culture of *remembering* the dead, or a culture of *caring* for the dead? Well, before blogs, smartphones and social networks, hardly anyone built a substantive digital legacy in life, and so the Internet was almost exclusively equipped to support the memory type of continuing bonds. Very few ordinary folks in the 1990s had a digital reflection or a digital shadow with any vitality or substance; you would be unlikely, logging on, to find a particularly vivid representation of anyone that mattered to you. If anything happened on the Internet in the 1990s to help continue bonds with the dead, it was through online memorials that were compiled post-mortem.

When Web 2.0 came along, and gigabytes and terabytes of data started being stored on personal devices and distant servers, the Internet began to house extensive archives of millions of ordinary lives. Technologically mediated 'remembering' became incredibly easy, because our modern technologies save data effortlessly, by default, silently uploading them to the cloud for safekeeping, and cataloguing them cleverly for easy search and retrieval – by us, or by someone else who knows the location and holds the keys. We went beyond using the digital sphere to *remember* the dead and started using it to *care* for them: to feel in contact, to connect with them. In essence, because the dead live on through tech, we're now witnessing a shift in how Westerners think about them. But let's go back to the beginning, when it was still all about memory.

A young engineer called Mike Kibbee came up with one of the first platforms to commemorate a large number of dead online, in one place. It was 1995, and Mike would soon be dead himself, from Hodgkin's lymphoma. He faced his impending demise boldly and pragmatically; first designing his own coffin, and then collaborating with a friend to create a technology that combined cemetery, eulogy and obituary, and translating them all into an online form. His own memorial on the site, which still exists today, replicates the obituary that appeared in print in the *Globe and Mail* in Toronto. The obituary describes Kibbee's idea for the World Wide Cemetery as 'a stroke of genius', enabling a far-distant son to 'visit' the grave of a parent through his computer. It further described how the online cemetery mirrored the mourning rituals of offline life, enabling visitors to online graves to leave digital flowers, a poem, or a condolence message. 'The wonderful interconnectivity of the Web', said the obituary, 'made it easy to link deaths (and the marvellous details of lives) of family members who may have died years apart and in different countries.'[97]

World Wide Cemetery touts itself as the oldest online cemetery and memorial site in the world. Just like offline burial sites, this 'elegant, peaceful and serene'[98] space is a separate, dedicated place for remembering the dead, and visitors can engage in all the familiar mourning rituals: sharing stories, memories and photographs, leaving virtual flowers, and writing messages to the deceased. And, just like in the offline world, some cemeteries flourish while some become disused. When the World Wide Cemetery was launched, being the first phenomenon of its kind, it garnered a considerable amount of media coverage, and there was good uptake. Most of the memorials on the WWC are for people who have little or no other online presence.

Marc Saner, who took over the running of the cemetery when its previous proprietor tired of the task, told me that he did so because he couldn't bear the thought of his godson having no accessible digital legacy. 'After his death it took only about one year to erase him from the face of the Internet,' he said to me. 'This bothered me. My WWC venture started with my desire to give my godson – and my father – a permanent online presence.' Permanent it may or may not be, for these days, the cemetery is an exceptionally quiet place. It has set aside a hundred-year fund to attempt to safeguard its existence over the longer term, taking it to 2095, but visiting it feels a little like going to a small churchyard where no one is buried any more. In 2017, as far as I can work out, one memorial appeared. At the time of writing, with the year a quarter gone, 2018 didn't look as if it had seen any at all. When I spoke directly with Marc, he was reluctant to put a number on new memorials. 'It's small,' he admitted.

Other digital cemeteries may be more successful because they make sense within a particular context, fulfil a current need, or even meet the agenda of the state. An online memorial site in Hong Kong, for example, is run by the government with the hopes that it will encourage cremation and other sustainable burial practices.[99] Someone can only create a memorial for their 'beloved one' if that beloved was cremated and/or interred in a Hong Kong cemetery or columbarium.[100] Legacy.com is an online memorial of a different kind, a flashy combination of search engine and death-specific news desk, a for-profit enterprise that allows people to access obituaries from more than 1,500 newspapers and 3,500 funeral homes across the United States, Canada, Australia, New Zealand, the United Kingdom, and Europe – usually for a limited time, in accordance with their terms of agreement with various newspapers.[101] It features tributes to celebrities penned by professional obituary writers on the Legacy staff, rather than by individuals who were close to the deceased.

In the virtual world Second Life, there are multiple memorial gardens, some dedicated to commemorating certain kinds of lives, or certain manners of death. On a rocky outcrop overlooking the sea, for example, with weeping willows and lights twinkling in the trees, is the Transgender Hate Crime and Suicide Memorial.[102] Elsewhere you can find the Peace Valley Pet Cemetery,[103] where there is no requirement for lost pets to have ever been a carbon-based life form: 'Walk through the graveyard and see the fun and touching ways Residents choose to memorialize their little lost loves, both real and virtual. In addition to the cemetary [sic], the build also comes complete with a chapel for delivering a fuzzy friend's last rites.' And if distance, schedule or infirmity prevents you from attending a funeral and paying your respects, you can often go to the funeral home's website to sign a virtual guest book.

All of these online facilities have much in common with their offline counterparts, while additionally doing what the Internet does best – making it possible for people scattered around the globe to share and access information and memories. Mourners should be able to engage in these rituals without let or hindrance, so it would be rare to be charged admission to a physical cemetery where the recent or contemporary dead are interred. Analogously, some dedicated memorial sites online are free to anyone with access to a browser. The World Wide Cemetery, for example, promises that they will never charge a fee or require accounts or passwords in order to visit the memorials. There are, however, many exceptions.

An erstwhile work colleague found out about the death of her distant friend's husband on Facebook. Overwhelmed with grief, his widow had simply posted a link to the obituary, without further comment. As we sat together in a coffee shop talking about this event, my colleague picked up her phone and scrolled back to her friend's original post, expecting it to be an expired link. 'Obviously now it's going to be different, because it was four years ago,' she said. She clicked on it anyway. To her surprise, it did link to the funeral home's obituary and Guest Book page, and there was his name – we'll call him Ralph Buxton. 'You can keep the memorial website for Ralph Buxton online by contributing to its extension', the page said. Alerting us to the fact that other 'commemorative products' were also available, it enumerated a schedule of the fees payable for a three-month, one-, two-, five- or ten-year stay of execution for the online memories of Ralph.

The ten-year option would set you back $499, Canadian. We sat in stunned silence for a moment. 'This Guest Book has been archived and is no longer available online,' we read. 'Restoring the Guest Book is a wonderful way to allow acquaintances and loved ones to express their sympathy and share fond memories.'[104] The site went on to urge even persons outside the family to consider restoring the Guest Book as a 'special gift' to Ralph Buxton's nearest and dearest. In this case, in a rather depressing instance of the commercialisation of death, keeping online memories 'alive' came at a price.

Online memorials of these types have a critical feature in common: the image of the person that they present is biographical rather than autobiographical, subject to the editorial control of the authors. The deceased person, a non-participant in this process, isn't around to challenge what anyone says about them. Mourners are at liberty to do what mourners always do, including smoothing out the rough edges, sanctifying or vilifying at will, and deciding what's most important to remember. Ultimately, the people left behind negotiate and establish a durable biography, a reasonably true-to-life picture of the dead person that feels comfortable enough for individuals or communities to carry forward,[105] and the deceased themselves won't have too much to say about it. When a person has built a substantial, visible digital legacy during their own lifetime, however, it's an entirely different kettle of fish.

Ω

Being at peace with the age that I am, I feel no shame in disclosing that I sit firmly in the 'digital immigrant' category, no matter how well I may have adapted to changing times. When I owned my first computers in the early 1980s (initially a Texas Instruments TI-99/4A, then a Commodore 64, followed by an Amiga), the most social interaction I derived from them was having my friends round to play Parsec. The World Wide Web hadn't been invented yet when I went away to university, but we had an intranet where I was studying, a space-age technological marvel that enabled students to chat one another up in real time via text messages, but that wasn't capable of revealing what the other person looked like, resulting in innumerable eventual disappointments.

After I graduated I went backpacking around Europe, during which time I effectively lost contact with all my friends and family for two months – it was before I had my first email address with America Online, and cafes at that time only featured coffee, not computers. When Friendster and MySpace hit the scene in 2002 and 2003 respectively, I was far too busy with graduate school to notice or care. My first experience of an online social networking platform, therefore, was Facebook. I liked it from the start, enjoying the ease with which I could post photos as well as updates, and keep track of my widespread friends and family. It was a boon to an expatriate who'd lived in a lot of cities and still missed some people who'd remained in those places.

And the first time I encountered death on a social networking site? About half an hour after I signed up, while entering the names of my high school and university friends to see if they were on this new platform too. Some of them I was actively interested in reconnecting meaningfully with, but for some I was just curious about what they looked like these days, what they were up to now. I typed in the name of one old acquaintance – we'll call her Jessica Smith – and a handful of Jessica Smiths appeared, most of whom looked considerably younger than the thirty-seven-year-old I was searching for. One entry, though, stood apart from the others. It wasn't because I recognised the young blonde woman in the photograph; it was because it said 'In Memory of Jessica Smith' next to the image. Curious, I clicked on it.

That was 2007, and it was only the year before that Facebook had leapt the confines of educational institutions to become available to anyone over the age of thirteen with an email address. Prior to its public launch, it is difficult to imagine that Mark Zuckerberg and his colleagues ever sat around a table saying, right, guys, eleven years from now we're likely to surpass two billion regular users worldwide. Millions of those users could die every decade. What are we going to do about that? How many people do we need on board to respond to any special situations that arise?

I wasn't there, but I'm willing to bet that the above discussion didn't occur. As you would expect from a site that was set up to connect living users rather than to act as a mixed-use social networking platform and massive digital cemetery, at first there was no facility to memorialise profiles on Facebook, and it would have been impossible to differentiate dead users' and live users' accounts on the basis of design features alone. You could only have been tipped off by the content on the Wall (as it was then known) and the sudden cessation of posts and other activity from the account holder.

Clearly, this was a situation that wouldn't work over the longer term, and it didn't take Facebook long to realise that they needed to act. In 2007, a year after going public, Facebook started putting the profiles of dead users into a 'Memorial State', initially employing a one-size-fits-all approach. Once someone had contacted Facebook about a user's death and that death had been confirmed, no one could log in or change anything about the profile again. All of the posts would revert to 'friends only', even the posts that had been public before. Notifications from the account, such as birthday reminders, would stop. People could still post on the Wall to memorialise and to communicate with the deceased but, other than this, profiles were frozen in time and looked the same.

Facebook didn't seem terribly sure about in-life profiles repurposed as memorial sites, though. 'In the Memorial State, certain profile sections and features are hidden from view to protect the privacy of the departed,' said a Facebook representative in 2007. 'We encourage users to utilize groups and group discussions to mourn and remember the deceased.'[106] So the page I had stumbled across wasn't the Facebook profile that Jessica had built and used in life. Instead, as the name suggests, it was a group started by her friends after she died, in some ways similar to something that could have been created on the World Wide Cemetery. I read the messages that people wrote and studied the photographs they posted, experiencing a sensation that I recognised, a kind of detached curiosity with a dash of poignancy. It was the sort of feeling you may have experienced when passing a place where a fatality has occurred, pausing to read the faded notes pinned to teddy bears, or tied to shrivelled bouquets of flowers and attached to a lamp post or railing – a slight voyeurism, mixed with a moment of contemplation on the fragility of life.

After a few moments clicking and scrolling, I started to wonder what had happened to Jessica's own profile, if indeed she had one. Facebook was pretty new; maybe she didn't have her own presence at all. With a couple of clicks, though, first on the name of the administrator of the group, then on the administrator's own list of Facebook friends, I found her. Not only did she have a Facebook page, but it was extremely well developed, and no wonder: twenty-one years old in 2006, born in 1985 in a technologically developed country, Jessica was a child of the computer age, a digital native. She and her friends had clearly been early adopters of Facebook, and their digital footprints had ranged wide and deep across the fledgling social networking site within only a few months. Her life was an open book: however risqué the outfit, however profane the language, however raucous the party, it was all there. She clearly lived life to the full, and she shared that life with a large audience, an audience that now included me.

Jessica's in-life profile was different from the memorial page. On the latter, people talked *about* her, but on the former they talked *to* her, carrying on conversations that had started in life. One photo showed her with her friends, lined up on lounge chairs in some tropical location. It had been posted before she died, and there was her own comment underneath: 'Oh my god, we look all skinny and tan!! I want to go back there!!!' After her death, her friends simply carried on the thread of conversation, as though it was the most natural thing in the world. The comment was casual, conversational, everyday. It was almost as though she wasn't dead at all, just away for a spell, not available for the annual jaunt to Florida this year, but maybe next time. I had worked with bereavement and grief in my clinical practice and studied it in my previous research, and there were lots of things related to mourning that sparked my curiosity about her profile page, but these are what captured my attention the most: the pedestrian, frequent conversations with her community of friends, persisting six months after her death, phrased in the second person and directed at Jessica herself.

The other thing I noticed was my own inner response. Reading Jessica's own words, following her conversations with her friends, viewing her images, and seeing her in the context of her relationships was a far different experience from looking at the memorial page. Trawling through her digital legacy, the one that she herself had contributed to, I started to feel a kind of familiarity, almost as though I knew her. In fact, I can still remember her face. I can still recall specific photographs on her page, even ten years later. That's not because I still look at the profile, though, for that's denied to me now. Later that year, when I was starting to formally research the subject, I looked for her profile but couldn't see it in search results. I didn't understand why, at the time, but now I realise that it had probably been memorialised and, in line with the policy at the time, rendered findable and viewable to friends only.

Losing track of that profile didn't bother me very much. I was just a researcher interested in a phenomenon, and I could always find other 'data' to use for my study; if not Jessica, then someone else. What I realise now, though, is that what was a neutral emotional experience for me, a minor irritation, would have been a completely different experience for someone else, someone who might have been close to Jessica but not on her Facebook friends list for whatever reason. What if her mother or her grandparents had been comforted by looking at her profile, and then woken up one day to find they couldn't see it any more?

I now have a massive digital footprint of my own. Facebook contains a personal archive – thousands of photos, videos, updates and commentaries – that shows who I am. My Instagram account shows what I find beautiful, my Pinterest posts indicate how I clothe my body and decorate my home, and Apple Music and Spotify records reveal the music that speaks to me most; but only a select few see all these. To the wider public, my blog discloses my thinking and opinions, my Twitter feed projects my professional identity and reveals my political opinions, and my website touts my therapeutic and writerly services to residents of greater London. In hundreds of thousands of words and thousands of images, I have written and continue to author my autobiography, telling the world who I am, what I do, and what and who I care about. This vivid, rich, multi-sensory record is almost completely analogous to my self-presentation in everyday life. Many years ago an academic in Korea described how digital being sits somewhere between the being of mind, *res cogitans*, and the being of body, *res extensa*, sharing qualities with both but qualitatively different.[107] Of all the elements of my digital footprint, my data on Facebook most fully capture my personality, values, humour, image, and the last decade of my history. What value might that have to those left behind, if I were to die tomorrow? If my profile disappeared after I died, my *res digitalis* would disappear along with it, and what might *that* event mean to someone? Potentially, quite a lot. Ava, a digital native who participated in one of my research studies, lost her close friend to an automobile accident. She succinctly summed up the vitality that her friend's Facebook profile held for her: '[If the profile were deleted] it would feel like I wouldn't be able to talk to her properly,' she said. 'It would be deleting the last bit of her that's still almost real.'[108]

Talk to her properly. I was hearing this sort of thing a lot, and it seemed that something more was going on than merely memorialising the dead. In the early 2000s, a study of mourners' communications on virtual memorials such as online cemeteries showed that only about 30 per cent of people talked *to* the dead.[109] Ten years later, I tabulated that of nearly 1,000 posts on the five Facebook memorial pages I studied, 77 per cent were addressing the person who'd died. It was pretty obvious when someone had more of a memory culture, when they held the view that the dead couldn't hear them. For them, writing a message was about showing sympathy and support to the family, not getting in touch with the dead. 'I know he can't read anything, what we are writing here,' said one such visitor, and it's interesting that they felt the need to emphasise this. 'But I just want to share my feelings with his friends and family.'[110]

On the other side, there were the remaining three quarters of posts. 'Even though it seems silly to talk through Facebook, I know u can see and understand every word I type,' someone says. 'I know u can read this, it just sux that u can't talk back . . . thanx for lettin me talk to u again,' says someone else. Facebook is apparently so exclusively effective at facilitating connection that an inability to log on could entirely prevent one from making contact. 'Happy late birthday! I did not have computer access yesterday . . . but I did remember your birthday and thought about you all day!' says one person. Another apologises as well, even though it wasn't about missing a birthday: 'I'm sorry that I haven't written to you for a while now, I know the castle is luxury but not so much that it has the Internet.'[111]

Is writing messages in this way just about obeying force of habit or following conversational convention, rather than being an accurate portrayal of what we believe about what happens and where people go after death? Or can we take these messages as a true reflection of writers' beliefs? Do people think that the dead are not only sentient, but actually reading their messages from some kind of Internet cafe in the afterworld? And might individuals' religious beliefs make a difference as to whether they believe that they're actually able to make contact? After all, we live in increasingly secular societies. In the ten-year gap between the 2001 and 2011 census in England and Wales, the percentage of people identifying as Christian fell from 71.7 per cent to 59.3 per cent, and the percentage of people reporting that they had no religion grew from 14.8 per cent to 25.1 per cent.[112] Since 1990, it's been reported that the percentage of Americans reporting no religious affiliation has swelled from 8 per cent to 22 per cent, nearly tripling in just under three decades.[113] In what may be a conservative prediction, it's thought that by 2020 there will be more Americans with no religion than there are Catholics, who made up 20 per cent of the US population in 2017,[114] and by 2035 Americans in the 'no religion' category may exceed Protestants, currently the largest denomination in the United States, with 45 per cent of residents identifying as some type of Protestant. With such a decline in religiosity, does this mean that the concept of afterlives and the idea of the sentient dead are diminishing too?

You might think so, but actually this isn't necessarily the case. Interestingly (and contrary to what I've always assumed), belief in God and belief in an afterlife don't always go hand in hand, and even atheists or agnostics may believe in some kind of life after death.[115] Although formal religious education, beliefs and affiliation have definitely all dipped significantly in the past decades, levels of belief in God and some kind of afterlife have remained high,[116] and people express their beliefs so readily in the online world that scholars are finding all this far easier to investigate.[117] Sociologist Tony Walter, who has conducted many studies using online material to better understand our current views about heaven, angels, souls and the afterlife, refers to the many 'surprising mentions of heaven in otherwise secular posts'.[118] In Sweden, according to one researcher, a spiritual take on death and afterlife is widespread and shares more in common with New Age thinking than traditional religion, and the life-after-death concept 'glorifies individuals and extols the bliss of an afterlife existence free from any punishment'.[119]

Walter has also observed a recent boom in references to the dead becoming angels, a notion that seems to be embraced even by many thoroughly secular people. An angel that used to be a human being, alive on earth, is a new sort of angel, completely unrelated to any sort of traditional church teaching; Walter thinks that belief in this sort of heavenly entity is a way of articulating relationships with the deceased, relationships that are facilitated by the online persistence of the now-angelic dead's data. 'The most common place where the angelic dead are encountered,' he writes, 'is online.'[120] There is a remarkable similarity between cyberspace and angels, he argues, in that people seem to see both as ways of travelling between earth and heaven. '[U]nlike souls locked up in heaven,' he writes, referring to how we used to think about the dead, 'angels can read social media posts. Thus technological development affords a new space for, if not creedal faith, then spiritual discourse.'[121] Whether the online environment is in itself a kind of new heaven, or just the medium for people to get in touch with the angels in some other sphere, evidence seems to suggest that both secular and religious people believe in the idea of the dead receiving our messages and looking out for our interests on earth.

This was borne out by my own research participants when I was studying the phenomenon of talking to the dead on Facebook,[122] a medium that Walter describes as 'a particularly amenable place for angels'.[123] I asked Ava, a secular young woman, if she felt as though Facebook was somehow different from, say, writing a letter and leaving it at her friend's gravesite. She didn't hesitate. 'I feel she will see it if it's on her [Facebook] wall,' she said. 'When I can't see what I wrote to her, I feel like she won't be able to see it too.' What about sending thoughts and prayers? 'You can think thoughts in your head, and think, "Oh, I'm hoping he can hear me",' said Ruby, who lost her cousin. 'But when you write something in Facebook, it's a more tangible way to communicate.'[124] Not just remember: communicate. What about going into the person's room, inhaling the scent of their clothes and being amongst their things, like Allem Halkic's father had done? 'It's strange,' said Clare, 'but part of me just feels like he sees it somehow. When I'm communicating with him on Facebook, there isn't that immediate reminder that he's gone. But when I see his name on his headstone in a silent cemetery or I see his room frozen in time, it's more in-your-face.'[125]

If angelic entities still have agency, then they still have social agency. In the majority of posthumous activity on memorialised Facebook profiles, and perhaps to a lesser extent on bespoke memorial pages, I witnessed plenty of evidence of this assumption about the dead. They're physically dead, yes. No one seemed in denial about that. But socially dead? Not just yet. Unlike the hungry ghosts of China, demanding that they get the proper attention at the proper times and in the proper forms *or else*, the dead on social networks are generally perceived as benign, and connections with them – as is usually the case with angels – are experienced as positive. Nevertheless, they are perceived to have a few expectations; or at the very least, you might still feel certain social responsibilities to them. They could be capable of feeling disappointed or neglected if you don't go online to wish them happy birthday, or you don't say hi for a while, so you should probably apologise if you've been neglectful. They're still interested in hearing everyday news about the football game, the birth of a baby, or that gig that you went to, so it's nice to go online to share it, just like those Japanese visitors to Itaru Sasaki's property, which I described in Chapter One, share the news of their lives down the phone line that leads nowhere, the wind telephone. They might feel hurt, abandoned or rejected if you de-friend them, so excising a deceased friend from your social network feels complicated and strange to many.[126] It's important to express your gratitude by going online to acknowledge and thank them for the messages they've sent and the help they've given you, although that's been via other channels – natural phenomena, for example, or intercessions to keep you from harm. 'Thanks for the dream you gave me, you weirdo.' 'There's been a really bright star in the sky lately and I know that that's you.' 'The car almost skidded over the median. Thank you for keeping me from going across all the way.'[127]

Might something about our digital environment actually encourage our belief in its ability to help us traverse earth and heaven, just like the angels, and cross the barrier between life and death? Whether we are secular or religious, whether we believe in an afterlife or not, why might we have an intuitive sense that communicating on a social network is an effective way of reaching across death's divide to our sentient, socially invested dead? Well, we're now accustomed to technologically mediated communication with remote others. We don't need auditory cues, don't require seeing someone's face, to feel confident that our message has landed. That confidence can be derived from two blue ticks in WhatsApp, your friend's icon sliding down to the last thing you said on Messenger, or an automatic 'read' receipt in your email inbox. To send a communication into the ether is to assume it has been nearly instantly received, no matter how far distant your correspondent lives.

Have you ever sent an email or an SMS to an unintended recipient? A message with sensitive content, perhaps? On those occasions, do you comfort yourself with the thought that maybe they just won't get it, just won't read it? Of course you don't. You *know* that they've received it, that they could be reading it right now, and you panic. If boundary-less, limitless, instantaneous communication is all you have ever known and that is your everyday experience of techno-logically mediated communication, is it really so odd that this sense would stop with your friend's death? Not being able to see or hear them has never stopped them getting your messages before.

There's an additional reason that the connection remains so strong on social networking sites, and you've already seen a clue – it's in something that Ava said. 'When I can't see what I wrote to her,' she said, 'I feel like she won't be able to see it, too.'[128] It sounds as though Ava is describing one type of internal continuing bond, a kind of merger or incorporation into herself, so that her friend is able to perceive things through Ava's eyes. But it also reflects something else, something fundamental to the nature of social networking sites. Their *raison d'être* is connecting people, positioning us all within an increasingly complex web of relationships, the algorithms suggesting new points of contact all the time.

The English translators of the German philosopher Martin Heidegger liberally sprinkle hyphens to convey our constant inter-connectedness with other people – we're never just Being, we're Being-in-the-world-with-others, and all of us Beings are never getting out from under that particular existential given.[129] Social networking sites are simply our Digital-Being-in-the-world-with-others. Sure, you set up the profile on your own in the first place, but from there onwards it's co-constructed, co-authored. When Mark Zuckerberg launched the new Timeline layout in 2011, he invited us all to write our autobiographies, generously offering his site as publisher. '[This is] an important next step to help you tell the story of your life . . . to highlight and curate all your stories so you can tell who you really are,' Zuckerberg said at the launch.[130] Autobiographers on social media, though, inevitably have a whole lot of collaborators. If you were writing a hardback book about your life, you could maintain total creative control and commit it to print: there it is, in black and white, from the horse's mouth. But on social media you write your autobiography together with your co-authors, your friends. If you interacted with a friend constantly on social media and you visit that profile after she's gone, it doesn't just feel like 'this was her'. It feels like 'this was *us*'. Or perhaps even 'This *is* us'.

Facebook has continued to evolve its approach to dead users, with the avowed purpose of making it easier for people to continue bonds. Much more will be said about this later in the book, but for now it's enough to know that one of their more recent innovations has been the Legacy Contact, a role akin to a platform-specific will executor, who is nominated by the user in life. A Legacy Contact cannot delete any friends that the deceased chose for themselves, but they *can* add someone who was not on the original friends list, like a parent or grandparent, maybe someone new to Facebook who's only joining to connect with the dead person's legacy. If that person feels funny about doing that (and there's research to suggest that people who are less familiar with social media tend to find interaction with digital remains unnerving rather than comforting),[131] the Legacy Contact can also download an archive of data, provided the deceased gave permission for that when they set up their legacy provisions. A downloaded archive makes it possible for people who don't have Facebook accounts to access the material.[132] It's also possible for users to stipulate that they want their profiles deleted after death, but for that to happen a Facebook user will have needed to have the awareness and motivation in life to delve into their settings and tick that box.

When you combine the fact that Facebook profiles are durable by default with the huge number of users (over two billion and growing), it's little wonder that the dead are proliferating on the world's most popular social networking platform. Just how rapidly the roster of the deceased is growing, or precisely when the dead will outnumber the living on the site, it's hard to say. The years 2065[133] and 2098[134] have both done the rounds in the popular press, but on closer investigation these seem to have been the products of fairly quick and dirty calculations. Certainly, accurate statistical projections of when the tipping point will occur are challenging to make, especially when we can't predict the peaks and troughs in Facebook's future fortunes, whether these will be exponentially amplified by savvy business decisions or derailed by major scandals. In the second quarter of 2018, for example, the company's growth stalled, perhaps due in part to controversies surrounding the 2016 US presidential elections.[135] Even so, researchers at the Oxford Internet Institute have done their utmost to produce a more thoroughly considered projection of the global accumulation of dead people on Facebook, using the best source material available: the demographic data of 1.9 billion current Facebook users, and population statistics from the United Nations. Even if uptake plunged through the floor and not one further person were to set up an account, they've calculated that 1.4 billion Facebook users would have passed away by the end of the twenty-first century. If the network were to continue riding the wave of success, on the other hand, racking up new users at an annual growth rate of 13 per cent, there could be more than 4.9 billion profiles of the dead on Facebook by the waning of the century.[136]

Whichever way you slice it, that's a lot of deceased users – the rumours of millions of dead folks hanging out and being rather less than social on social media are entirely true. While the World Wide Cemetery never posed much of a threat to physical burial places, Facebook has become the largest 'cemetery' in the world. Offline cemeteries and monument companies, doubtless wondering whether their own coffers will steadily dwindle to nothing, see Facebook as the competition. Many have decided that if you can't beat 'em, you might as well join 'em. David Quiring's Seattle-based monumental masonry company, for example, is one of several companies producing QR codes that can be embedded in traditional granite headstones. Their website explicitly positions their QR-to-website service as an alternative or adjunct to a memorial on Facebook, suggesting that those left behind could compile obituaries, information about family heritage, photos and comments on a Living History™ archive website.[137] John Troyer of the Centre for Death and

Society envisions a 'Future Cemetery' that you can walk through wearing a virtual-reality headset, encountering your reanimated ancestors along the way,[138] and I've seen demonstrations of 'augmented reality' technologies enabling photographs to pop up on the screen like Pokémon when you aim your iPad's camera at a perfectly ordinary-looking headstone.[139]

All these technologies could certainly make physical cemeteries more engaging and meaningful places but, as the years pass, will we continue to be motivated to visit actual places of interment, when instead we could easily continue bonds out of the rain, on our phones, in the comfort of our own homes? But cemeteries with no visitors aren't the only concern here, because just as digital remains and digital technologies afford the potential to continue bonds, they can disrupt them too. So let's look at the other side, through a couple of stories that sound as though they would never actually happen.

Ω

First, call to mind your nearest, dearest, most lifelong friend. Ideally this is a childhood companion, someone you met at school perhaps, whom you have known and cherished for as long as you can remember. It could be a friendship of more recent vintage, but key to this exercise is that this person is integral to your life. If they had not been there for you, and there with you, you would be a far different person today. No one knows you like they do, and in your time you've generated a lot of memories together. You couldn't possibly count how many conversations you have had over the years, how many messages you have sent one another, or how many photographs you've accumulated together. Imagine that you keep all these relics of your friendship in a large shoebox, with your friend's name written on the lid. It's a large box, because everything is in there – the origami-folded notes you passed under the teacher's nose, letters in their stamped envelopes from when you moved to separate cities, and stacks of photos that span the decades.

One day you get the worst news that you could ever receive: your friend has died suddenly, in an accident. You are numb with shock and disbelief. They have always been there – a part of you has died with them. Grief-stricken, and aching for a sense of connection with them, you rummage in the top of your closet, bring down the shoebox, and place it by your bed. In the days and weeks after the death, you pore over the box's contents many times a day. It's not the same as having your friend back, but everything in that box is redolent of not just who your friend was, but who you were together. Its contents make you laugh, cry, smile. Kind messages flow in from friends and family who know of your terrible loss, and you place these in the shoebox as well. After some months, you notice that you are spending less time looking through the box, but that's all right – you know where it is if you need it, if the impulse takes you, if you want to find that thread of connection again. The box is just there, within arm's reach.

One day you hear a knock at your front door. Opening it, you are surprised to see your friend's parents, whom you have not seen since the funeral, and who have travelled many hundreds of miles to arrive at your front door unannounced. Silently, and without even acknowledging your presence, they push past you and walk down the hallway to your bedroom. They unceremoniously bundle up the shoebox and, as abruptly and wordlessly as they arrived, slam their car door and drive away. You are overcome, frantic; you are desperate to retrieve your memories. When you are finally able to contact these people about their inexplicable behaviour and the whereabouts of your shoebox, they have terrible news: they have destroyed the box, and all its contents. Your blood thrums in your ears, then feels as though it's drained from your body. You can almost feel your friend's hand slipping from the grasp of your own. You've lost them all over again: it's like a second death.

This time, instead of being the friend, you are the parent – the parent, say, of a teenaged son or daughter. One day they go out as usual; several hours later, the police are knocking at your door. Reeling with grief, you begin to do what you must: you pull yourself together to inform everyone of what has happened and to make the necessary arrangements. As a parent, you naturally assume that you will have a central role in arranging the celebration of your child's life, an occasion where everyone will come together, remember, and support one another in mourning this profound loss. When you take the difficult step of picking up the phone to call your closest family and friends, however, you are floored to discover that they already know, that they heard about it virtually as soon as it happened. Some of your child's friends have far more details about the accident than you have yourself. The news isn't yours to tell.

Although you're rattled, somehow you manage to resume your planning. You search for photographs to illustrate your child's life, but you realise that you don't have any recent ones. Once again, you get in touch with your child's friends for help. To your astonishment, you learn that all the photographs they have are currently in use at a memorial service for your child, a memorial that is taking place right now. It's being attended by hundreds of people, and it's being held in a location where you've never been before. You rush to get there, but you find the gates locked. Beyond the closed doors, you can hear people laughing and crying together – they are showing photographs, sharing stories, supporting one another and talking about your child. You knock, but no one answers. You pound louder and kick at the door, pleading for access and threatening to burn the place down if you don't get it. That's your child. You should have been a part of this. You should have organised this. At the very least, you should be there.

But no one lets you in.

Ω

If you're thinking that these scenarios are absurd, you're right. Anyone who enters someone's home without invitation to steal and destroy a box of important documents could be arrested and charged with trespass, theft and wanton destruction of property. And what sort of people would band together to plan and attend a memorial service for a deceased friend that deliberately excluded the friend's parents? But, in essence, this is what's happening. When digital remains cause a disruption in continuing bonds, it's almost always about one or both of these things: access, and control.

There are three broad categories of troubling experiences around *access*, all of which can interfere with a mourner's capacity to determine for themselves the kind of continuing bonds they want, or don't want. The first is the threat of the digital remains vanishing entirely, for ever. Mechanisms like automatic backups in the cloud may lull us into a sense of security about the persistence of all those data, but don't be fooled. Digital beings have an anxiety-provoking, paradoxical dual nature. On one hand, they can stick around for ever, and on the other they can vanish without a trace.[140] Because there are no more shoeboxes in the attic, the fear of losing access to a loved one's digital remains is intense, perceived as a second death, a thief in the night that can steal the last hope of connection. 'I would be close to inconsolable,' said one of my own young research participants. 'Having something that may seem so small to some people is everything to me. [His profile] is the one last thread of him that I have. If we lost it, it would be like losing him all over again. There are just certain things that rip the wounds open.'[141]

Your connection to someone's ongoing digital 'self' can be severed for ever through a host of mechanisms: a software glitch, an expired webpage subscription, terms and conditions of a particular platform that stipulate deletion on the account holder's death, hardware malfunction or obsolescence, or the next of kin removing a social networking profile. If you're a digital native whose communication history and photographic memories are stored completely on Facebook, your best friend dies, and his or her profile is deactivated: someone has entered your house, stolen your shoebox, and destroyed every last memory in it. And it's all perfectly legal.

The second problem with access occurs when you know the digital remains are out there, and you would love to get to them, but you can't. While in typical circumstances it is unlikely that anyone would ever lock you out of an offline memorial service, and no one could ever bar you from laying flowers at a headstone in a public cemetery, you could easily find yourself in a situation where your loved one's persistent digital reflection, perhaps the most comprehensive representation of them that survives in this world, lies frustratingly near but beyond your reach. Perhaps you weren't a Facebook user, but your child was, and they left no Legacy Contact that could add you. You would hope that someone could sit down with you, take you through it, and screenshot or download the bits you want; on the other hand, people may exclude parents or others from their social media for good reason, and loyal friends may be aware of that. In addition to exclusion from *material*, however, you'll also be excluded from the community of mourners that exists in that place. One mother, a participant in another researcher's work,[142] describes the devastation of precisely this kind of exclusion. When her son died, she said, she was surprised to receive no cards, no letters of condolence from any of her son's friends. During the funeral, she was told by his best friend that there was a Facebook page memorialising her son, but without a Facebook account of her own, she couldn't get to it. All she had, she said, was an 'empty mantle', the place where she would have placed cards containing words of solace. She didn't know what she could do about it.[143]

This issue can disproportionately affect family, rather than friends. Once upon a time, when someone died, it was the family, the legal next of kin, that had privileged access to the dead: to their bodies, to information about them, to their personal effects, to their writings and photographs. Law of succession was everything. Having privileged access also meant having control over who *else* had access, and family members were the keepers of that gate. In contrast, friends were at the opposite end of the access spectrum. However important their relationships might have been with the dead person, friends were in danger of being left out, of becoming disenfranchised mourners – a term that describes those whose grief is unseen, their right to mourn unrecognised or unacknowledged.[144] The long-time secret mistress who wanted her letters back would likely not have got past the front porch. School friends might never hear of the death or might not know in time about the memorial ceremony that they would have wished to attend. In an extremely short time, the balance of power has flipped, for the era of social networking is the Age of the Friend. If a mistress is on Facebook and Instagram, she'll have far more access to the digital legacy than a grieving but technophobic wife. A father might certainly find himself with no physical photographs, appealing to friends for images to display at the funeral. Scenario number two doesn't seem so absurd any more, does it?

The third issue with access is when there's too *much* of it. Now that memorialisation of profiles is more widely utilised, it's somewhat less common to encounter digital presences that 'act alive' – unmemorialised profiles suggesting that you wish your dead friend Leslie a happy birthday or propose that you like Glamping Holidays Inc. because your dead friend Keith quite enjoys them too. Profiles that are unmemorialised also act alive through being accessed and managed by others, with consequences that may prove difficult for many mourners. Some people, through technical ignorance or psychological naivety, have no idea how much their activity on a dead person's site can discombobulate others, but sometimes they're aware and proceed anyway.

A heart-wrenching illustration of this latter scenario is the story of Vanessa Nicolson, who derived comfort from being on her late daughter's unmemorialised Facebook account, which was still logged in as though Rosa were still alive. One day, Vanessa saw a significant status update from Adam, Rosa's boyfriend at the time of her death. The post made clear that he had started a new romantic relationship. Writing in the *Guardian*, Vanessa explained what happened next.[145] The feeling that her daughter had been 'replaced' gave Vanessa excruciating emotional pain. Almost without conscious thought, Vanessa posted a comment underneath Adam's status update. 'Don't you think it's a bit soon to fall in love again?' she said. As soon as she had done the deed, she snapped back to her senses, and realised that it would appear on the stream of comments underneath the post as a message from Rosa, Adam's dead girlfriend, complete with her profile photograph. Panic was added to the horrible cocktail of emotions Vanessa was feeling in that moment, and the impact on Adam and other friends is hard to imagine.

Indeed – unsurprisingly – when a spouse or parent is controlling the profile, research indicates that it has an almost universally negative effect on other mourners' ability to cope.[146] Even if a profile *is* memorialised, however, with no danger of startling messages popping up from beyond the grave, it could still prove problematic. There's no separate cemetery on Facebook, one that you can choose to visit or not – although the design for memorialised profiles is always evolving, at the time of writing, even memorialised 'Remembering' profiles are integrated within the list of living friends, listed alphabetically with everyone else. This makes oscillation between the dual processing model's 'loss orientation' and 'restoration orientation' as simple as a click or a swipe – living friend, dead friend, and back again. It's like encountering tombstones dotted here and there on the busy streets of a bustling community. The possibility of connecting with a Remembering profile is always there, and it may not always be easy to decide: is this a good time for me to stop, remember and connect, or a good time to just keep walking?

Ω

The other potential disruption of continuing bonds derives from challenges to *control*. Even if you do have access to part or all of a person's digital legacy, there may be troubling things about it that take your agency away, and that disrupt your ability to continue a connection. Problems with lack of control come in myriad forms, and one is death notification. Social media transmit news more quickly than any police force on earth, so just like in the second scenario above, news of a death may appear on Facebook or Twitter before immediate family members know about the death themselves, before they have any opportunity to process the loss, or before they can make decisions about disclosing it. Those notifications can then form part of the enduring footprint, a perpetual reminder of the pain of that moment of realisation.

Encountering unexpected bits of a digital legacy can also upend what you thought you knew about someone, shredding the integrity of the bond to pieces. You can imagine, for instance, what her grieving husband might have felt if he'd stumbled on Searcher #711391's search history, full of her sexual fantasies, her plans for clandestine meetings with other partners, and her dreams of leaving him to go to Alaska. Even in the case of a digital legacy that is largely positive and a source of comfort to a griever, the presence of upsetting or traumatic elements can taint the whole picture. The profile photo might be all wrong. A family might not even be able to change a profile that is plastered throughout with depictions of the perpetrator(s) responsible for their loved one's suffering and death: a woman killed by her partner, for example, or someone who ended their life after being abused or bullied by a friend or family member. Less dramatic, but just as problematic for continuing bonds, the digital footprint might not constitute the kind of durable biography that an individual mourner can live with – and there may be nothing they can do about that. Any time that multiple parties with disparate needs and preferences are accessing a central set of digital remains, it's likely that not everyone will be happy with all that appears there. Who gets the last word on who the dead person 'really' was?

Sometime in the early 1970s, Aidan decided to leave the UK and start a new life.[147] In so doing, he left behind a dependent family, including a six-year-old daughter, Cathy, and that family struggled on in poverty after his departure. 'He was an alcoholic wife beater who deserted his children and left my mum pretty much destitute,' says Cathy. She never forgave him for leaving, and she never saw him in person again. Some years later, however, something discomfiting happened – she received a Facebook friend request from him. Having never considered her errant biological dad a father at all, much less a 'friend', Cathy ignored the request. Aidan persisted. Eventually, Cathy relented. What if he were planning to return to the UK for a visit, or for good? She was from a relatively small town, and she wouldn't want to be surprised by running into him in the high street. What if he had changed? And even if he hadn't, what if he died? Even if he didn't play any role in her life, shouldn't she know?

Eventually, Aidan did die, but even though they'd never actually interacted on the site, his digital presence there proved difficult for Cathy. One of her siblings had reconciled with their father, much to Cathy's displeasure, and the people who had known him in more recent years used his memorialised profile to glorify him. 'There's this profile picture that I've always hated of him with my [siblings' kids],' Cathy explained, 'that's never going to go away. And it's just always there. There are all these people who are sainting this man who is so far from being ... I'm not saying he didn't have good qualities. But history's just been completely rewritten in this page that just *sits* there, and there's nothing I can do about it. And I can't de-friend him, because then I won't ever be able to see what's going on ... but at the same moment I can't bear to look at it, because it just kind of makes me vomit when I read all this stuff that isn't *him*.' Aware that she was in the minority opinion, she was loath to post her own version of events on her father's Facebook profile, and she is still grappling with her feelings months after his death.

Ω

Like Robert Louis Stevenson's famous character, Aidan was both Dr Jekyll and Mr Hyde. Both personae were equally true, at least to the people that knew each of them. For the American members of Aidan's family and social circle (according to Cathy, who confronted some of them via Messenger about what she perceived as their hurtfully inaccurate or at least incomplete portrayals), his persistent digital self was a good fit with the man they knew. His digital remains facilitated their own bond with the father, husband and friend they'd lost. For Cathy, Aidan's enduring biography was nothing short of a travesty, a constant niggle and source of pain, a thorn in the side in her grief process. 'It's not complicated grief,' she insists – she's familiar with the terminology of bereavement – 'but it's complicated.' Complicated, indeed. When a digital legacy is a salve for some and a wound for others, what should happen to it? Who decides? Well, more often than you might expect, it's corporations. To illustrate that, let me tell you the strange and sad story of Hollie Gazzard.

Chapter Three
Terms and Conditions

It was 18 February 2014, and twenty-year-old Hollie Gazzard was probably anticipating some much-needed normalcy at the hair and beauty salon where she worked. A few days prior, she'd managed to finish her relationship with Asher Maslin, a troubled and violent young man who had coercively controlled her throughout their relationship. He had always become particularly upset when faced with the possibility of her leaving, so it is unlikely to have been an easy week for her. But hairdressing was her abiding passion, and after the break-up, a regular day at Fringe Benefits and La Bella Beauty should have been a respite for her. She was nearing the end of her shift when her ex-boyfriend entered the salon with a knife. Despite strenuous efforts by emergency services to save her life, Hollie died from fourteen stab wounds.

There were multiple reasons for the young woman's murder to garner widespread media attention. Hollie had reported Asher's threats to the police, and there were questions about whether someone could have proactively intervened to save her life. In addition, this was a highly unusual incident in Gloucester, a relatively small city with a low crime rate. The chief inspector of police in Gloucester pleaded with members of the public to bring any video footage of the incident directly to police, rather than posting it online, for apparently a number of witnesses had filmed the attack on their smartphones.[148] At first, the danger of videos spreading online, potentially negatively influencing the investigation and trial, was the most pressing concern around social media that law enforcement – or Hollie's family – needed to think about.

After Asher was convicted and sentenced to life in prison a few months later, however, it didn't take long for Hollie's bereaved family to become aware of surprising and upsetting things emerging online. First came the trolls. While we may sometimes assume that the Internet created the phenomenon of trolling the dead, saying terrible things about deceased people is hardly new. Documents dating to the second century AD identify Chilon of Sparta as the philosopher who coined the maxim, 'Don't speak ill of the dead.'[149] That was around 600 BC, and he would hardly have needed to offer such advice unless folks were actually going around doing it. But while human beings are capable of unfathomable unpleasantness both online and off, the digital environment has made the impact of bullies and trolls even nastier, enabling cruel comments to stick around, allowing them to spread far and wide, and rendering them more difficult to erase and hence forget.

I met Hollie's father, Nick, in a cafe just a few doors down from the salon where his daughter had been murdered. Nick appeared in Chapter One, extolling the comfort that Hollie's digital legacy gives him, and describing how it brings home the 'joy of Hollie' to anyone who encounters it. Very soon after the death, however, some elements of unauthorised biography cropped up that were anything but joyous. As soon as Hollie was named in the press as the victim, trolls picked up the scent of a wounded family, and they wasted no time in mounting their attack. Nick remembers being alerted by a friend that they'd been targeted. 'Someone sent us a link, saying, look, we've seen this, you need to know about it,' Nick explained. 'Internet sites in the US were berating me for letting my daughter go out with a black man. A particular website in the States had pictures of me, pictures of my granddaughter, of [Hollie's sister] Chloe, of Hollie, of my wife.' Eventually the trolls became bored with the lack of response and redoubled their efforts in ways that were more personal and direct. 'Then there were trolls on Facebook as well, on family sites. It was a very similar thing – about letting your daughters go out with a black boy. Bearing in mind Chloe's partner is also mixed race, as are my grandchildren, it was quite alarming. Very distressing.'

'What kind of time frame are we talking about?' I asked. 'After you lost her, when did you become aware of problematic things happening online?'

'Oh,' said Nick. 'Within weeks.'

Still reeling as they were from Hollie's murder, the Gazzards weren't sure what to do. How could you control what strangers were saying online? Where do you start? 'It was happening not only within the UK, but across the world as well . . . these negative things coming in, which we're having to try and deal with,' Nick said. But they *had* to do something – it was too horrible to endure. Somehow they gathered the presence of mind to get help, to seek solutions. 'We tried to get it shut down via GCHQ [Government Communications Headquarters], the FBI, that route,' Nick said. 'The police were very good on that front, they took it on board. I said, this is what we need to do. Could you do it? Please? And they said, leave it with us. We'll deal with it. And . . . they dealt with it. I think. It did get taken down in the end . . . not 100 per cent sure.' He paused. 'I never then went back through it. Because it was something I didn't want to go into.'

Later, I have a look myself. I feel a bit sick to my stomach, like I'm doing something dirty, as I type in Hollie's and Chloe's names, alongside the cruel keywords that the tormentors used in their assault on Hollie's memory. It appears that, at the very least, the links between Hollie and the troll websites have been broken. At the bottom of the search results is a blurb that says, 'Some results may have been removed under data protection law in Europe.' Nick assumed that the Gazzards had expressed their concern to the proper authorities, that the authorities had listened and, true to their word, they had dealt with it. 'So we had to contend with that,' Nick said to me. 'One upset.' He held up an index finger, showing that what he actually meant was, 'upset number one'.

At the time Hollie died, Nick was a self-professed Web 2.0 novice. He had email, of course, but he'd never used any social media platforms. Facebook was so off his radar that even now Nick wildly underestimates how widespread it was at the time that Hollie died, seeming to assume that in 2014 most people's awareness level about Facebook mirrored his own. 'Facebook was still fairly early. We're talking nearly four years ago now, it wasn't that well known,' he said, referring to a period when two out of three of Facebook's then 1.28 billion users logged on every day, with Americans devoting fully one fifth of their smartphone usage time to checking the app.[150] No wonder he was startled at what they saw, when Chloe was introducing Nick to Facebook after Hollie died. 'Somebody had set up a "RIP Hollie Gazzard" site,' he said, referring to a memorial group much like the one I'd initially stumbled across, when I first encountered the intersection of death and the digital. 'And it instantly got something like 15,000 likes.' I asked who set it up. Was it a friend, a family member?

'We didn't know him,' said Nick. 'In one way it was pleasing, because people had thought about her, but the second thing was ... actually, we had no control over it. Anything could be going on there, being said. So it was a bit alarming, bearing in mind we'd already had [the trolling].'

By this time – which was quite a while later, Nick said – the family had started their own Facebook group in Hollie's memory, associated with the charity that they'd established, the Hollie Gazzard Trust. Having this other Facebook group hanging around wasn't just worrying because of the risk of trolling, it was potentially confusing; Nick thought that many people might mistake it for an official page, run by the family. He wondered if it might even divert attention away from the charity. The family didn't want to lose the likes and comments on that site, but having already been burned so badly, they couldn't help being worried about their lack of control. Once again, they made a proactive bid to seize back the reins. Chloe contacted the administrator of RIP Hollie Gazzard, Nick explained, and proposed that the Gazzard family manage it instead.

I held my breath for the next part of the story, remembering an anecdote that one of my research participants, Clare, had related about what had happened when her cousin died. In that case too, multiple competing memory groups had sprung up on Facebook. 'People are eager to jump on opportunities, just for the popularity,' Clare said. 'One guy, who is always trying to be in control of every-thing, was doing it just to get the credit. And my friend overheard another girl complaining the week after the accident, because people were leaving *her* in-memory-of group to join mine. It's just so petty and immature.'[151]

But the administrator of RIP Hollie Gazzard wasn't petty or immature, and the result of the family's efforts was positive. 'He came back and said, yeah. Here's all the details,' Nick said. 'So Chloe then managed it. I'm not sure whether it's actually still going now because we put everything to the main trust site.' I remarked that they were fortunate. The administrator of this group could easily have missed or ignored Chloe's message, or even refused the request. 'Yes,' said Nick, 'and then started posting things. It wasn't setting up a fake account, you know, it was "RIP Hollie Gazzard". He wasn't trying to be anyone else. But it just goes to show what can be done. If he [had refused to hand over control, and he] hadn't been malicious or broken [Facebook's] terms and conditions, there would have been nothing we could have done about it.'

I was unsurprised that the administrator of RIP Hollie Gazzard was so amenable to handing over control to Hollie's family. He was a stranger, unconnected to the family, but something about Hollie's death had moved him to create a forum to memorialise her. That's not particularly odd; like the previous chapter points out, in individualistic cultures we often choose whom we commemorate on the grounds of emotional resonance rather than family ties, and sometimes we elect to honour people we've never met. Many months on, however, it seems unlikely that an unconnected person would have been monitoring the group particularly closely, and Nick was right – anything could have been happening on the site. If the administrator hadn't replied and handed over the login details, they would have been precluded from doing anything about it; but, as it was, their fears were instantly assuaged. They were free to merge the content of that page with the Hollie Gazzard Trust page, or not, as they saw fit – they were back in control. Upset number three, however, made the first two look like a walk in the park.

Nick might have been a social media novice in 2014, but Hollie certainly wasn't. She used it all – Twitter, Instagram, Facebook. After Nick joined Facebook himself, after her death, he was struck at what he discovered by visiting his daughter's profile. 'It was a whole new world,' Nick said. 'Her interaction with lots of people, friends, family, who were also on Facebook. That was great. It was our contact with Hollie, really. It felt like her telling me, these are the things I've been doing.'

The fact that Nick could see Hollie's posts at all was a bit of a lucky break, and it was only possible for two reasons. First, Facebook had recently made a crucial change in its policy. Between 2007 and early 2014, once the profile of a dead user was placed into a memorial state, all of the posts on that profile would be set to friends only. People signing up to the network for the first time would never have been able to access a dead user's memorialised profile, because *nothing* would be public, and the deceased wouldn't be around to accept new friend requests. Those people who hadn't been on the deceased's friends list, but who had been able to see public posts before, suddenly wouldn't be able to see them any more. Just a few days after Hollie's murder, however, Facebook announced in a press release that the company had been doing a lot of thinking about how they could best respect the privacy of their deceased users, as well as showing the maximum amount of sensitivity to the bereaved. 'How might people feel? Are we honoring the wishes and legacy of the person who passed away? Are we serving people who are grieving the loss of a loved one as best we can?'[152] They concluded that the best way of striking this balance was to keep people's privacy settings in death exactly the same as they had been in life. In some ways, Facebook would be keeping the same contractual privacy pact with their erstwhile account holders as they'd had when those users were alive.

If Hollie had been a more guarded or cautious type of person, her privacy settings would likely have been more stringent. Had 'friends only' been her default, Nick wouldn't have been able to access her profile at all – at least not through his own, newly forged account. But that simply wasn't Hollie's personality. She embraced any opportunity to connect with friends old and new, and there were 700 of them on her friends list. The vast majority of her posts were visible to anyone. 'Yes, most of it was public,' Nick said. 'Probably she didn't understand all the settings. And secondly, she would have wanted everyone to see it anyway, because she was like that. She lived her life in the social arena, and we know that because of the amount of distress it caused people when she was killed.' In line with the new policy, the visibility of Hollie's posts had been retained as they were, set in virtual stone, and Nick had an untrammelled view of virtually everything Hollie had ever posted. And 'everything', of course, included multiple images of Hollie with her murderer.

$$\Omega$$

When stories about this situation first appeared in the news, the number of photos mentioned was nine. Being unfamiliar with the structure and design of Facebook, including its various photo albums, Nick initially thought that this was how many there were. Later, he saw that there were about eight times that many: seventy-two. The fact that the man who killed Hollie was plastered all over her social media isn't surprising. In fact, it was probably statistically *likely*. Hollie was a twenty-year-old woman. In January 2017, the Pew Research Center reported that 88 per cent of people in the United States between the ages of eighteen and twenty-nine were using Facebook, and that 69 per cent of women overall were signed up on the site.[153] Add in the Centers for Disease Control and Prevention (CDC) statistics indicating that women are far more likely than men to be killed by someone that they know, and that more than half of female homicide victims are killed by intimate partners.[154] In the United Kingdom, where Hollie lived and died, 44 per cent of female murder victims in the twelve-month period ending March 2015 were killed by partners or ex-partners.[155] Finally, consider how likely it is that a relationship will be visually documented on the autobiographical stream that is someone's social media feed; in January of 2018 it was reported that there were 350 million photographs per day uploaded to Facebook.[156] When you combine the likelihood of a relationship's being documented on Facebook or Instagram, with the likelihood of being murdered by someone who is close to you, it becomes highly probable that, if you were to fall victim to this unfortunate fate, your social media legacy would depict you alongside your killer.

While some people choose to purge their accounts of photos of their ex after a relationship ends, people's social media practices after a breakup vary widely, and of course Hollie and Asher had only recently separated. Was leaving them there a preference, a deliberate choice, something that reflected her wishes? Was it an oversight? Did she delete some but miss those seventy-two? Did she intend to remove them, only not to have the chance before that horrible afternoon when Asher appeared at her workplace? We'll never know whether Hollie was on the cusp of taking those photographs down, or what their continuing to be there meant to her. What is clear, however, is what they meant to her family.

As far as her other social networks were concerned, Nick hadn't really looked into or thought about her Instagram much, and he was vague when I asked him about Twitter, assuming that it had been deleted since she didn't tweet any more. He was incorrect about that – at time of writing, four years after her last post, it was still there. For Nick, at least, Facebook was the most particularly compelling part of Hollie's digital legacy, but the comfort he felt when he saw her profile was overwhelmed by the upset caused by photos of Asher. 'It makes me feel sick when I look at those photos,' her father said, speaking to a BBC journalist in 2015, 'and to be truthful I try not to go into her Facebook, as I get quite distressed by it.'[157] When we met, he described how the photos would 'bang you in the face' when you opened up her page. I notice that in the hour and a half I sat with Hollie's father, even when he referred to the man who killed her, he never once spoke his name.

Since they'd been able to handle the first and second upsets, with some help, the Gazzards assumed that they'd be able to tackle this new problem with relative ease. Thoroughly modern estate advisors often counsel people to keep records of their login details for their next of kin, but Hollie left no virtual key-ring with the necessary tools to unlock all the different areas of her digital legacy. Even so, the family didn't think that this would be an issue. Hollie's sister, Chloe, knew her password, so they could simply use it to access the account and remove the offending photos. Couldn't they? But there was a problem. By the time they tried to log in (Nick thinks that this was a few months after she died, although his memory is fuzzy), the profile had been memorialised. Nick's not sure how it happened, because the request didn't come from anyone in the family. 'We can only assume that with the media publicity, [Facebook had] picked it up,' Nick shrugged, 'maybe done searches or whatever, and then they'd memorialised the account. They did it off their own back, Facebook.' Logging in wasn't possible any more, so the fact that Chloe knew the password didn't make any difference at all.

Even though they'd hit that barrier, the Gazzards thought it would be a straightforward process to get the photos selectively edited out of her profile. Surely if they contacted the social media site and explained the situation, Facebook would understand that the photographs were too upsetting, that they should be deleted. 'So I wrote to them and said this is the situation, we'd like a number of pictures removed from Hollie's site. Can you give us access to the site, so we can manage it? I don't think we got any response.' But the Gazzards weren't prepared to give up. 'We emailed again, and I think we got a response saying . . . OK, we can't change it, but what we can do is shut Hollie's site down for you,' said Nick. 'No no no, that's not what we want! We don't want the site shut down. We wanted certain things removed. That's so insensitive. I can go to the bank and say, can you shut that account down, because she's no longer here, and actually, that money is due to me, because I'm the beneficiary of her estate. Fine. Go to Facebook . . . no. It's a complete blank.'

So they tried another tack. The family had made numerous contacts in the media after Hollie's death, and one journalist asked Nick to help out with a piece about digital afterlife. Also contributing to that story was a solicitor called Gary Rycroft, a wills and probate specialist with a particular interest in digital assets and digital legacy, and he had some advice for them. I'd seen Gary's name in news stories about the Gazzards, and I rang him to ask about his recollections. 'Facebook were being very black and white,' Gary said to me. 'Either close down the account or memorialise it, leave it as is, but you can't edit it now – it's one or the other. Nick and the family were saying, we don't want to close it down, we want to visit the site, we want to post messages to her . . . so we don't want to do that but we don't want to see *him*. I said to Nick, well look, there is a middle way. Hollie took the photos of him, and she owned the copyright. That's now with you as her executors and beneficiaries, and you say you withdraw the copyright to those pictures. That's what Nick took to Facebook, I believe, and that's what they agreed to.'

I wasn't sure how to respond when Gary told me this, because at that point I'd already had my cafe lunch with Nick, and my understanding was somewhat different. The Gazzard family might have eventually made headway by undertaking a copyright claim directly with Facebook, but it never came to that. Help came instead from an entirely different and unexpected source. The Gazzards had collected about 11,000 signatures on a petition and had renewed their contacts in the local media, such as BBC Points West in Bristol. 'It was just [about] keeping the profile up there,' said Nick. 'What could we do?' It was then that they were contacted by the Web Sheriff.

Before he received a cold call from John Giacobbi,[158] Nick had never heard of Web Sheriff, and neither had I. It appears that the Web Sheriff patrols the Wild West of the Internet, engaging in showdowns with villains like Internet pirates, intellectual property thieves and privacy invaders. Giacobbi, the CEO and founder, seems to be an intellectual property lawyer who deals with celebrities, film studios, record labels and rock stars. The Web Sheriff website is emblazoned with graphics of old-fashioned star-shaped sheriffs' badges and gets mentioned in places like *Rolling Stone*, the *Hollywood Reporter* and *Billboard*. 'A rock 'n' roll client once said to me that keeping the internet clear of piracy and other issues was like weeding the garden – you have to keep on top of it or the weeds grow back; but, if you keep at it, you'll end up with a beautiful rose garden,' says Giacobbi on his website. 'In many ways, this is a very accurate analogy. Many are the times when we've been called and asked to help transform a "netscape" from weeds to roses.'[159]

Certainly, Asher was a weed in the rose garden of Hollie's own netscape, but wasn't Giacobbi's company just set up to serve celebrities, actors and musicians? Not necessarily. 'These days, you don't have to be a Hollywood A-Lister or a rock 'n' roll superstar to have internet issues or online problems,' Giacobbi says in his CEO's message on the website. 'Everyone from small businesses to online shoppers, families, schools, and even children – who can be particularly vulnerable – are at risk from all manner of cyber issues, including e-fraud, impersonation, online libels, invasions of privacy, trolling, cyberstalking, and worse.'[160] It's hard to imagine how much worse it can get than having the image of your daughter's murderer all over her enduring digital legacy. When he heard about Hollie's story in the news, Giacobbi took it on himself to get in touch with Nick.[161]

'He said, look, Nick, I've found out what's been happening. You're having a problem with Facebook . . . . I will get these removed for you, and any other quotes, photos that you want taken down across the Internet, Google or whatever . . . I'll get them removed,' Nick recalled. 'And he did. And he got some stuff taken down on Google, so when you put searches in certain things don't come up.'

'Did you just go on one day and find that they were gone?' I asked.

'John let us know,' Nick said. 'He said, it's all sorted. It was two or three weeks later. They'd all gone.' Finally, her family and friends could engage with this digital legacy and experience the joy of Hollie.

Ω

On one level, I look at the Gazzards' story through a psychological lens, noticing how their problems with access to and control over Hollie's digital legacy complicated and compounded their pain and suffering. As a psychologist, I know how lack of control, and an inability to predict what will happen, can contribute to stress and even trauma. Had the Gazzards possessed the foresight, skills and head space to anticipate and address the problems that were arising online, and had they had more power over Hollie's digital legacy throughout, their pain could have been reduced. But control and predictability weren't possible, for a couple of reasons. The first, of course, was simply to do with the nature of the technology. In enabling information to be rapidly proliferated far and wide, the Internet enables us to affect and be affected by far-distant strangers, for good or for ill. Without the Internet, the trolls in America would probably never have found out about what happened to Hollie in England. Without social media, they wouldn't have been able to target her family with racist diatribes that came out of the blue. And the other reason for the family's lack of power? Law. Whenever laws that were forged in a pre-digital age collide with modern information technologies, a general state of confusion and contradiction tends to hold sway, but when death is added to the mix, you've got a mess on your hands.

It's not a revelation that laws, regulations and their enforcement can have considerable emotional impact, but I've been somewhat surprised to realise that it's virtually impossible to understand the psychology of bereavement in the digital age without also getting my head around the law. Early in my research and writing process, having witnessed that laws often contributed to the pain and suffering of bereavement online, I emailed the London-based Centre for the Study of Emotion & Law,[162] asking whether they were studying anything to do with death and the digital. A very short email came back to say that nope, it hadn't been a focus thus far. I replied that it really should be, for it seemed critical to me that the emotional impact of this area is acknowledged. So I guess that's what I'm trying to do here. Luckily, I know a handful of super-specialised legal minds, people who've spent a lot of time in the murky region where law, death and the digital meet.

In 2012, I collaborated with several of these lawyers, their specialities varying but all interested in death, for the Panel on Death and Post-Mortem Privacy at the Amsterdam Privacy Conference.[163] This was the event where I first encountered the film *I Love Alaska*, which presented the inadvertent autobiography of AOL Searcher #711391. It was a fascinating presentation, if I say so myself, delivered to a rapt audience of about three people. Whether conference-goers were put off by the word 'post-mortem', doubted its relevance to privacy, or couldn't find the room, I'm not sure. For me, though, participating in such a panel helped me realise for the first time just how many strands of law are involved here: contract law, privacy law, intellectual property law, and succession law. As you've probably already worked out, this is not a tidy, easily negotiated intersection where everyone knows the rules governing traffic. It's a poorly signed, confusing roundabout that bereaved families may find themselves circling for years, a place where accidents happen and people get hurt, a place where there is no universally agreed highway code. And, importantly, it's a place where even the people who planned and built the streets struggle to guide us on how we should navigate them. The best guidance we've got comes in the form of the terms and conditions that govern our use of any given Internet site – and there's your first problem.

Before I get into how terms and conditions matter when we're dead, let's think about T&Cs in more general terms. In the pre-digital era, the terms and conditions that affected our lives the most were probably the ones with a direct financial impact. On more than one occasion, in a state of high dudgeon, I've rung up a mobile phone, credit card or utility company to protest an extortionate fee or unexpected charge, confidently assuming that they'd made a mistake. In nearly every one of those cases, I was put in my place by a kind but firm customer service assistant referring to the terms and conditions I'd agreed to when I contracted to use their services. Banging down the phone, I would vow to read the T&Cs more carefully next time . . . and then I wouldn't. For online services, the T&Cs we tend to care about involve how our data will be used and our privacy safeguarded, and this is such a prime consideration for the online world that 'privacy policy' is nearly synonymous with 'T&Cs'. Even if you're using a free online service (and sites like Instagram, Facebook and the basic versions of some email services are currently all technically free), you can't assume that there will be no costs of failing to scrutinise the terms of use.

Back in 2008, two researchers from Carnegie Mellon set out to estimate some of the costs – in time and money – of thoroughly reading the privacy policies of all the sites we use and visit online. Not how much time we *do* devote to this task, mind you, but how much time we *could* devote to a responsibility that is generally seen to lie with us as consumers. 'Under the notion of industry self-regulation, consumers should visit websites, read privacy policies, and choose which websites offer the best privacy protections,' Cranor and McDonald wrote.[164] I'm guessing that undertaking such a scrupulous process for each of the websites you visit sounds rather unfamiliar to you, and if Cranor & McDonald's calculations are anywhere near correct, our antipathy to such a task is understandable. The median length of a privacy policy at the time these researchers conducted their analysis was 2,518 words; based on an average reading speed for 'academic' writing of 250 words per minute, that adds up to about ten minutes per policy. It doesn't sound like much, until you consider how many websites you visit over a twelve-month period. Around 2007/2008, when they undertook their study, Cranor and McDonald estimated that the average American visited 1,452 sites each year. Nationally, this resulted in a $781 billion hypothetical opportunity cost for America annually. On an individual level, it would take 244 hours a year to read all the terms and conditions of the sites you visit, and double that to additionally take the 'comparison shopping' approach Cranor & McDonald spoke about, adding up to about seventy-six working days a year.

I don't know about you, but I'd struggle to devote seventy-six working days to reading T&Cs, much less the additional days that have doubtless been bolted on in the ensuing decade, during which smartphones and mobile devices have virtually become extensions of the human hand. Lack of time isn't the only reason we don't read T&Cs, of course. First of all, like Hollie Gazzard, we might not care. 'She would have wanted everyone to see it ... because she was like that,' said her father. But he also surmised – probably correctly – that she wouldn't have understood all of the settings anyway.

'I did a little one-off kind of project with some teenagers ... explaining to them in plain English what the terms and conditions were of some sites like Snapchat and Instagram,' said Gary Rycroft, the solicitor who helped the Gazzards. '[I said], did you know when you sign up that these companies can read your private messages, we can sell your data? And these kids were saying, I didn't know that! If I'd known that, I wouldn't have done it! Teenagers and grownups don't understand that when they tick the box where you agree to those terms and conditions.' It's also rather difficult to comprehend something we can't find. I'm reminded of the passage in *The Hitchhiker's Guide to the Galaxy*, when Arthur Dent is lying in front of a yellow bulldozer, vehemently protesting the imminent demolition of his house to make way for a bypass. An unpleasant man from the council counters that the notice of demolition has been available in the planning department for some time – did Mr Dent not find it? Arthur splutters that he certainly did, but that it had been rather difficult, given that it had been on display for any interested residents 'in the bottom of a locked filing cabinet stuck in a disused lavatory with a sign on the door saying Beware of the Leopard'.[165]

Just as Douglas Adams might have predicted, often T&Cs are timed less than optimally, or placed several clicks away from where we're actually *using* a website or app, or otherwise thoroughly hidden away in the furthest reaches of a site. There are things that could be done about this. Research shows that we're far less likely to clock terms and conditions when they're shown in the app store, and more likely to notice them if they're visible to us when we're actually using the app.[166] But even if we can find them, we might not be able to understand the arcane legalese, meet the university-level comprehension level required, or interpret the vagueness of the phraseology used. Perhaps motivated more by compliance with regulation than by a moral obligation to consumers, T&Cs have grown more complicated rather than less, longer rather than shorter.[167] We're too overwhelmed with information and preoccupied with instant gratification to inform ourselves thoroughly, even when that information pertains to things we genuinely care about. While it's possible that Hollie herself might not have cared much about data protection, many social media users do, and in March of 2018 thousands of Facebook users were alarmed and outraged at news reports that the site's security had been 'breached' and the data of millions of users 'hacked' by the allegedly nefarious Cambridge Analytica.[168]

The red mist of anger dissipated somewhat when it was further clarified that Cambridge Analytica had only been able to acquire personal data because the persons in question – Facebook users – had given *permission* for third parties to access their data.[169] That said, much like the council in *The Hitchhiker's Guide*, Facebook hadn't always made it easy for users to review and understand their choices and to exercise control over their data, so in this case, as in so many others, it wasn't entirely the fault of the users. Fortunately, various countries around the world are taking steps to address the problem of lengthy, complicated, hidden T&Cs. In the European Union, for example, the new General Data Protection Regulation (GDPR) came into force in May 2018, which required all organisations processing data to provide information about how they do this, in a 'concise, transparent, intelligible and easily accessible form, using clear and plain language'.[170] Great news – although I still notice a creeping fatigue as I page through all of the new, improved information about how companies are processing my information. But anyway, now that regulations that govern our data are evolving, we should shortly expect to be far clearer about lots of things, including what happens to our data when we're dead . . . right?

Well, not so fast. In one of the GDPR's briefer clauses, it explains that the new regulation 'does not apply to the personal data of deceased persons. Member States may provide for rules regarding the processing of personal data of deceased persons.'[171] It's not particularly surprising that the EU shuffled the responsibility for this confusing area of data protection to the member states, but I can understand why they would feel justified: data protection doesn't usually deal with the dead. The laws covering data protection, privacy, copyright, and contracts generally only cover 'natural persons', that is, living persons with legal status. Natural persons enjoy human rights; they can own physical and intellectual property;

they can enter into and maintain legally binding contracts; they can expect their data to be appropriately protected in line with accepted standards. The dead can't do any of that and, not being natural persons, they have no such rights.

While the rights and obligations of some types of contract might pass to the deceased's heirs, the accounts that individuals hold with online service providers aren't like that. Most online services include a one account/one user policy as an integral part of their terms of use, and one can understand why. The invisibility cloak of the Internet, the inability to identify individuals online via their physical and facial characteristics (thus far, anyway), makes all manner of criminal activity possible – identity theft, impersonation, copyright violations, bank fraud, malicious communications, defamation, libel. When an Internet company can't be sure who its users are, it's poised on the edge of a liability snake pit. It seems to make sense that when someone dies, the contract should dissolve, and the account be made non-transferable, even if its contents can somehow be passed on. Nick Gazzard thinks that it's unlikely that such considerations ever troubled Hollie's mind. 'Hollie would never have known what she was signing up to on Facebook and how that would affect her when she was gone,' said Nick. 'She wouldn't have thought about it, even. Even if it had been in black and white. You're invincible when you're that age.'

It isn't just users that are confused about these things, however. It's the companies that control our data. They may have great power, but where the death of users is concerned, they're struggling to accept and respond to the idea that they also have great responsibility. When I talk to Gary Rycroft, the solicitor who advised the Gazzards, I tell him the story about Kim's frustrated attempts to close down her father's Amazon seller's account. 'That's just a sign of the fact that it's a new area, and companies haven't got their protocols together,' he said. 'There isn't a legal requirement for companies to have these protocols, but I think there will be a commercial imperative at some point, because it's a service that you're giving to your clients. I think as people encounter problems and complain about it, and as companies get bogged down, they'll think, OK, guys, we have an issue, it's taking time; we need to make it slicker. I think there will be a commercial reality there.'

Perhaps, I thought, I should give companies like Amazon the benefit of the doubt. After all, it had been a year and a half since Kim had first contacted them about her father's death. Perhaps in the meantime they had got to better grips with these commercial realities. So I decided to set myself a task – first I would try to work out, from Amazon's own website, what a bereaved person should do with their loved one's account. I set the timer on my watch, but gave up after twenty minutes, having gone down all manner of dead ends, no pun intended. Then I rang customer services.

'Could I ask, if someone has died and had a seller's account on Amazon, how would I go about closing their account?' I asked.

'You want to close down an account?' said the customer service representative.

'No, no,' I said, 'I don't want to close down a particular account. I was just looking on your website to see what the procedure was if you *did* need to do such a thing, and I wonder if you could point me to where I could find that information.'

I was put on hold for about five minutes.

'There's actually no information on the website,' he said, when he returned. 'But all you'd need to do is pass security and request closure of the account.'

'What would that entail?' I asked.

'You'd have to verify the person's phone number, their name, and the email associated with the account,' he said, although he didn't seem at all certain.

'Wouldn't you need the person's password?' I asked.

'No, you wouldn't need the password,' he said, sounding even less convinced. He seemed to get more confused when I pointed out that with a phone, name, and email address alone, you could shut down anyone else's account if you wanted, irrespective of whether they were dead, and that this didn't seem right. And wouldn't you need proof that the person had passed away?

'Um, no,' he said, sounding more rattled all the time. 'You wouldn't need that. Just the phone number, name and email address. So you want me to escalate this to a supervisor?'

I did. When the hapless customer service representative came back, the information was different. This time, he said that if you possessed the password you could just log in and request closure, but actually if the person were dead you would need to send in a copy of the death certificate and that then it would be shut down. I suggested that it might be useful to have this information somewhere on the website, so that bereaved persons could access it more easily. He didn't have much to say on this, so I asked who else I could speak to about this idea and was put on hold for five further minutes. When he returned, he said that someone from the 'Internal Reviews Department' would get in touch with me shortly. I'm still waiting.

Throughout this conversation,[172] I was acutely aware that I was merely a writer who was digging around, getting in touch about a hypothetical scenario. I had experienced no loss, was in the midst of no grief. Every time I was put on hold, my empathy for Kim increased threefold. I thought of all the other incidents I'd heard about, tales of family members who'd experienced maddening or insensitive encounters with data controllers; particularly tales told by next of kin, inner-circle mourners who, like the Gazzards, assumed that they would have far more power than they actually did.

Ω

The phrase 'law of the land' implies a particular locality – this is how we do things around *here*. Just as we have local cultures of grief and mourning, we have local laws about death and succession. Just as there are many lands, there are many variations on those laws. Some types of law are relatively consistent from land to land, such as copyright law, for which there is a generally accepted convention worldwide. Data protection law is a bit more variable, with pockets of convergence and lines of divergence. European data protection law may be quite different from the United States, but *within* the

European Union, it's the same all over. If you reside or have your business in an EU member state, you're going to be subject to the General Data Protection Regulation. But the laws that govern how assets are passed on when we die? On the spectrum ranging from the least to most harmonised type of law, they're at the jangling, discordant pole.

Dr Edina Harbinja is one of my aforementioned super-specialised law buddies, and she let me know straight off the bat that in our relatively short conversation we wouldn't even begin to be able to cover the specifics. 'We can't even look at every jurisdiction, we'll just look at principles,' she said. 'We can't look at twenty-seven or twenty-six [EU] member states, and we can't look at all the US states and analyse those.' Briefly, she explained, in the United States, laws of succession fall under state jurisdiction, not federal, so Nevada may be different from New Hampshire. In the European Union too, succession laws vary significantly from one member state to the next. A person in Hungary and a person in Finland may fall under the same data protection and privacy laws, but they may have wildly different regulations about succession. 'Succession laws are really, really different,' she said, shaking her head.

We only had a couple of hours, but I had a *lot* of questions. I realised it was complex, but I wanted to at least try to get my head around a handful of things. In the UK and the US at least, I wondered, do digital 'things' even count as things that can be passed down, as personal property that is executable in a will? You hear lots of talk about digital wills, but would one hold up in court if it were challenged? Do Facebook Legacy Contacts and Google Inactive Account Managers (IAMs) have the same authority as will executors? And what is likely to happen if, like Hollie Gazzard, you die unexpectedly, without a will, a digital will, or someone designated to carry out either one?

Unlike Hollie, I do have a will, which is still relatively unusual, even for someone in their forties. I wrote it up myself using a service I found online, rather than utilising the advice of a wills and probate solicitor, and I assumed that I could leave whatever I wanted to whomever I wanted. I reckoned that I could design everything to happen exactly the way I wanted it to, from disposition of my estate and my body to the songs played at the funeral, and that my will would set that in stone. It's understandable that I would have assumed this. In many parts of the Eastern world, in more collectivist cultures where family is emphasised over the individual, laws often dictate that you *must* leave certain proportions of your wealth to certain people. But in the Western tradition of personal autonomy, the tradition I come from, testamentary freedom is privileged. So my assumption that I could make up my own mind about which beneficiaries would receive my personal property was correct. My assumption that pretty much everything came under the definition of 'personal property' – digital things included – was not. While Edina was explaining the laws that dictate what we can and can't do with traditional physical and intellectual property and assets, everything seemed relatively straightforward, manageable and clear, even accounting for the differences between countries. However, once we added the word 'digital' into the mix (digital property, digital assets, digital works), things started to get interesting.

I asked Edina about the case of Justin Ellsworth, the soldier who died when he was hit with a roadside bomb in Fallujah. Like Hollie, he died intestate, with no will, digital or otherwise. Like Hollie, it was highly likely that he never fully read the terms covering his use of the platform and didn't know what would happen to his data if he died. The cases were different in one significant way – Hollie's digital legacy was publicly visible, while Justin's case involved private emails. However, unlike Hollie's father, Justin's father was able to fight against a service provider's T&Cs, and the service provider delivered what he was after: access to his child's personal data. As described in Chapter One, John Ellsworth was able to procure a downloaded archive of the entire contents of Justin's Yahoo! account, even though Yahoo!'s terms and conditions had stated clearly that the account should be closed on the death of the account holder. Justin's dad had argued that, as next of kin under the laws of succession in California, he had the right to this material.

I maintain a broadly pro-privacy stance, even where the dead are concerned, but to try and understand this situation, I played devil's advocate with Edina about the controversial court decision to release Justin's private correspondence to his father. Not so long ago, I pointed out, financial and personal records and communications were all in paper form. There was no alternative. If you wanted your personal papers free from prying eyes in the event of your death, you secured or destroyed them as you went along, or made provisions in your will for a trusted executor to dispose of them (which one could apparently do in the UK, because they satisfy the tangibility requirement even if they're technically worth nothing). Paper is tangible; boxes of correspondence are tangible. For anyone who died intestate, without having destroyed their private correspondence, those left behind would be free to read whatever they liked, subject only to the dictates of their own conscience. So even though they weren't made of paper, weren't Justin's emails really just the same sort of thing as paper letters? Wasn't the court right to think of the contents of Justin's account as being akin to traditional property, that would pass to a person's next of kin if that person died without a will? If John Ellsworth, as his dad, could have easily accessed Justin's correspondence and financial information in a more analogue era, the court might have thought, why shouldn't he have the same ability now – just as so many other grieving parents feel they should?

'From the "black letter law" perspective – the rules as they are – it can't be seen as property,' said Edina. 'It doesn't satisfy the requirements of property tangibility. But besides that, online information is highly *personal*. It's personal data, personality, *identity* . . . . They are so different from the offline property that is not that personal: traditional property, land, wealth, financial assets.' But physical letters can be personal, I argued. They can be imbued with the personality and identity of the writer, and *they* can pass on to next of kin as physical property.

'There is this element of physicality in the letters, yes,' Edina said. 'But also it's less information, it's less data, it's less complicated. There is no account there that is owned by a company. It's a piece of paper, a physical possession, a letter. The person owns it. So it's more simple offline. Online, you have the intermediary service provider, multiple individuals involved, data, servers. The account comprises many different elements.'

So digital 'stuff' may be considered differently to non-digital 'stuff' in terms of tangibility, and it may or may not be considered legally executable, depending on the laws of the land. But what about your intellectual property? A person's intellectual property, intangible creations of your mind that may have value, and that can be stored in a variety of formats, *do* pass to one's next of kin. Edina explained that yes, one's heirs are entitled to copyright for a certain period of time, depending on the system – it could be fifty years or eighty years after a death. John Ellsworth couldn't use copyright as a basis to procure Justin's correspondence, since his emails weren't communicated to the public or seen as original creative works. But Hollie's posts on Facebook, including photographs she took, were almost always set to public. Before the Web Sheriff succeeded in removing Asher's images from her Facebook profile, Hollie's next of kin had hoped to use the copyright argument to get photos – particularly photos that Hollie herself had taken – removed from the site. Maybe that was possible, because Hollie's profile was so public. Indeed, maybe that's the route the Web Sheriff took. But what if Hollie had only had ten friends, or fifty, or a hundred? Would it have counted as 'published' then? And what about me? As a keen photographer, I definitely count my carefully crafted images as my intellectual property. If I were to post one of those images on Facebook, with my invitation-only viewership of 300-odd people and my Alcatraz-level privacy settings, does that count as 'published'? Even though they couldn't log into my account, would my next of kin be able to argue ownership and have say-so over it if I died?

'It's not clear,' said Edina. 'There is a lack of case law. For Facebook, it depends on your privacy settings. So, if you're setting something friends only, or friends-of-friends, it's a limited circle. I would argue that it's not communication to the public. It's personal data, it's private, it hasn't been published as such, so copyright in published works is not in play there ... but yeah. Definitely if your settings on Facebook are public, then it's communication to public and intellectual property is relevant there. The problem is that there are so many issues around it. Intellectual property, data protection, property, contract. And it's a *mess*. It's a mess. It's not regulated.'

When I'm speaking with Edina about this hopelessly tangled legal landscape, I feel a charge of intellectual fascination. When I speak to Michaela, I just feel sad.[173] Michaela's story, situated at the blurry boundaries of the physical and digital, caught somewhere between tangible property and intangible creative/intellectual property, illustrates that it's not only *online* data that is contested when people die. Michaela's long-term boyfriend Luka was a photographer and filmographer who stored precious still and moving images of the couple's trip to California on his laptop, not backed up anywhere. When Luka died suddenly, in the flat the couple shared, his family took possession of Luka's things, to include the laptop.

So far, that situation sounds pretty clear. According to UK law, the laptop was a valuable, tangible item belonging to the deceased person, who died intestate; as the couple weren't married, his parents were next of kin. His photographs and films, Luka's intellectual property, also passed to his next of kin under laws of copyright and succession. At first it didn't matter. Michaela wasn't in the best place emotionally, and she had more pressing concerns. She needed to look after herself. One day some years later, though, missing Luka more intensely than usual and wanting to see the images again, she messaged his brother on Facebook and asked if she could get a copy of the data on the laptop. The response she received was brusque and somewhat defensive. Couldn't she have asked for this stuff years ago? The laptop had been broken for ages. Michaela was devastated. She'd always assumed that she'd be able to have a copy of the material when she was ready, and apparently it was gone for ever.

But, as with all these matters in the digital age, the complexity didn't end there, because it wasn't just Luka's creative output that was on the computer. Michaela is a photographer too. Because she'd used Luka's camera to take pictures of Luka, these had also been stored on the laptop. Clearly, it wasn't the case that absolutely everything was gone – certain images had been downloaded and kept by family members before the laptop broke; images with huge personal and sentimental significance to Michaela. She knows that because the family posts them on Facebook now and again.

'It's really strange when they post photos on social media that are photos that *I* took,' she said. 'That was my eye. That was my eye, seeing him. They're my photographs.' I ask, tentatively, whether she'd thought about pursuing a copyright claim, which would seem to have a good chance of succeeding. But although Michaela found these posts painful, she wasn't interested in taking it up legally. She'd tried to accept it by doing other things, like planting a tree in his memory in a nearby park. 'Things that are about growth,' she said, 'about the future.'

Ω

Even though the photos of Luka that crop up on Facebook are a source of conflicting emotions for Michaela, the situation she confronted wasn't so much about a corporation. It was about a conflict between friends and family about who's entitled to look after the remains of someone who's died, and their possessions. More on that in a later chapter, but back in the world of the data that we hand over to online service providers, many of the corporations that control and process our data in life clearly have a fair amount of catching up to do where death is concerned. Others have made an effort to start cutting through a mess that's so sloppy and confusing that Edina and I got through a fair few cups of coffee talking about it.

The first to take a major step forward was Google, in 2013, when it introduced its Inactive Account Manager (IAM) feature. In contrast to my experience with Amazon's website and call centre, I'm able to discover details about Inactive Account Manager quite quickly, so I'm initially impressed by the transparency as well as the way in which users appear able to tailor what people can and cannot access. Google users nominate a trusted contact, who won't get a message when they first set up an IAM, but who will receive a notification when they have been inactive on anything Google-related for a specified period of time. 'John Doe (john.doe@gmail.com) instructed Google to send you this mail automatically after John stopped using his account. Sincerely, The Google Accounts Team', reads the exemplar text for such a notification, which the IAM would receive after the longest possible time a citizen of the modern age could be expected to abstain from generating a Google-activity ping. The user decides whether the IAM has access to all of their data or just a portion of it: 'John Doe has given you access to the following account data: +1s, Blogger, Drive, Mail, YouTube. Download John's data here.'[174]

Clear and reasonable, I think, but when I try it for myself – using the link and trying to log on with my own Gmail account – a cartoon image pops up. It depicts a distressed-looking robot holding a spanner, with both of its legs and its left arm in bits on the floor. 'The setting you are looking for is not available for your account,' says the caption underneath. Several clicks later, I'm still not able to work out whether it's really true that my particular Google account doesn't entitle me to determine my data's fate after I die and, if so, why. Even if I never figure it out and never nominate an IAM, my next of kin just might be able to get hold of my data anyway. Who knows? 'None of the providers ever promise, in the terms and conditions, we *will* do this,' said Edina. 'So Google, if you didn't use Inactive Account Manager, they have a policy to allow access to some of the content, but they say they won't guarantee this.'

Does Facebook's Legacy Contact offer any more of a guarantee? That I *have* been able to work out, for some time now. This feature, introduced in 2015, seems less fine-grained than Google's IAM but perhaps more straightforward to set up. As described earlier in the book, Legacy Contacts have limited powers. While a Google IAM could see emails if the dead person had arranged for it, a Legacy Contact on Facebook can't see private messages. Although the editorial and moderation powers of legacy contacts have since expanded[175], at the time I spoke to Edina in autumn 2017, their capabilities were still extremely limited. The downloadable archive was restricted to material from the publicly visible Timeline. Legacy contacts could only add friends, not delete any that the deceased person themselves had added. They couldn't change the deceased's own privacy settings, and in fact couldn't change much at all apart from the lead photos and the pegged post at the top of the Timeline. I was comfortable with someone else having that relatively non-interventionist, non-invasive involvement with my Facebook profile after my demise, so I had happily nominated my partner as my legacy contact, assuming that this decision would hold weight. My partner, however, wasn't yet my next of kin, and I wasn't sure of the law in the UK, so I had a question for Edina.

'Tell me what would happen,' I said. 'Let's say somebody's got a Legacy Contact. It's not their next of kin, it's a friend, or whatever. And let's say that the next of kin is the one who gets in touch with Facebook and they say, here we are, we're next of kin, we're informing you of the death of our son or daughter. They know nothing about the Legacy Contact, and they say, we want this taken down. Presumably Facebook says to them, so sorry, Mum and Dad, so-and-so identified a Legacy Contact, and that person is in charge of managing this and there is no stipulation that this be deleted after death. So that always would hold up—'

'No,' interrupted Edina, emphatically. No? 'Not in the UK,' Edina reiterated, confidently. I was a bit shocked. So that's definitive? I asked. 'That's for sure,' said Edina.

I noticed that this was the most flustered I'd felt during the entire discussion. I had recently been on multiple television and radio programmes assuring the British public that the Legacy Contact feature was a way to seize control over what happened to your digital legacy, and I was wondering just how much of an idiot I'd made of myself. 'So . . . so . . . in the UK,' I spluttered, 'somebody could nominate or identify a Legacy Contact, and feel secure in the knowledge – um, their assumption, rather – that that person would have the power to preserve that account in the way that they wish it to be preserved. But the next of kin could still come along and . . . and . . .'

'The courts wouldn't recognise that,' said Edina, matter of factly. 'A will has formalities. It's not digital, for now. It can't be electronic. It has to be signed, it has to be deposited with the court and executed as the law stipulates. So an electronic will, or any version of electronic disposition, is not recognised in the UK at all.' Discomfited, I changed the subject and asked about the situation in the United States. Edina explained that the United States boasts the 2015 Revised Uniform Fiduciary Access to Digital Assets Act (RUFADAA)[176]. The 'Revised' bit is there because a previous version of the Act (the UFADAA of 2014) had decreed that digital stuff should be treated like physical stuff, and held that executors and other fiduciaries were entitled to comparatively *carte blanche* permission to access a deceased person's online accounts. In other words, under UFADAA the material in online accounts was seen as conceptually equivalent to a box of letters in the attic. Unsurprisingly, this stance prompted howls of protest from the big-tech 'custodians' of this digital material – Apple, Facebook, Google and the like – as well as from privacy advocates.[177]

RUFADAA was a compromise, allowing executors and other fiduciaries access to certain content as long as the deceased had affirmatively authorised it.[178] According to Edina, having this degree of clarity about digital material enshrined in law was revolutionary. 'It's the first time that this has been recognised anywhere in the world, where they say explicitly, if the deceased has stipulated online, using any online service or technology, that they want certain digital assets to be disposed of in a certain way, that will override the will for that particular asset,' she said.

By July 2019, nearly all 50 states had made RUFADAA law in their particular bit of the land.[179] So, if you're a US resident and you've appointed a legacy contact on Facebook or an Inactive Account Manager on Google, you can be pretty sure that your wishes will hold legal weight. But what if you live outside the United States? Do you know whether your arrangements for your digital legacy are legally enforceable where *you* live? I thought not.

So, what seemed relatively simple didn't seem at all simple any more. What I thought I had control over, I don't. Dr John Troyer regularly teaches a module on the sociology of death at the University of Bath. To each new group of students, he puts a handful of questions. One is, 'If you could pass on just one digital thing to your descendants, what would it be?'[180] For me, the answer to that is my Facebook profile

– the publicly visible contents, that is. As a writer, a photographer and an expatriate, I have used that platform for over ten years to portray my life, my passions, my happy and sad events and my opinions to distant friends and family, and to record things of sentimental value for posterity. It's not only richly autobiographical, but it also charts the development of my young daughter's life, her personality, her appearance. I was shaken that my arrangements for this treasured digital artefact did not carry the weight I thought they did, and astounded that I had never realised all this before. I'd been studiously immersed in this topic for years, a supposed expert . . . not a lawyer, but someone who plays one on TV, and in this chapter. If I didn't know about this, what hope did anyone *else* have? And when can we expect our laws to catch up with the new digital realities?

Just as Gary Rycroft said that commercial incentives would eventually drive companies to sort it out, the reality is that only when big problems arise and are tested in courts of law everywhere, we will begin to pick out a new set of laws that are fit for a digital era – a digital era that's so different that an entirely new set of laws may need to be created, rather than modifying the old ones. Edina, for whom this is an ongoing area of scholarship, went a bit wistful when she talked about it. 'Unfortunately for me,' she said longingly, 'we haven't had a big fat lawsuit about this. When we do, they'll realise just how messy it is. It would open up more discussion.'

Without a critical mass of those 'big fat' court cases, things proceed at a stately pace, about 1/100th the speed of the ongoing digital revolution. In the United Kingdom in 2017, the Law Commission launched a public consultation with the entirely laudable aim of modernising the law around wills.[181] So far so good, until you get to the bit about digital assets, the bit where they said they weren't going to regulate those for the time being. That seemed crazy, but then I understand why so many lawmakers don't want to touch it. Most of the people involved in reforming the law remain firmly in the digital-immigrant camp. 'I don't want to be disparaging towards my fellow professionals, because they all, you know, most of them try and do a good job and do the best for their clients,' said

Gary Rycroft. 'This is an emerging area. I think, I'm sure that not all of them have a great level of knowledge, because I'm thought to be someone that knows quite a bit about this area, and I still feel like a bit of a novice myself sometimes, so it's inevitable that other people are a bit further behind.'

I know the feeling. I myself am an immigrant masquerading as a native, but I spend my time immersed in the subject of death and the digital. I understand its importance. The lawmakers have less expertise, wider priorities. They work in law and government; I work with the bereaved and with the practitioners who help them, witnessing first hand the human consequences of the current lack of clarity. I understand how much more painful grieving can become when bereaved families lock horns with corporations in the legal Wild West of digital legacies. Happily, even in the face of rigid terms and conditions, and confusing laws and policies, solutions to individual problems can sometimes be found, as in the Gazzard case. It's the best solution that could have been achieved, but it's probably not the typical story. Nick wishes that things could change so that no more families would have to go through what the Gazzards endured.

'Certainly more awareness needs to be made for people who have these accounts as to what will happen to them once they are gone,' Nick said. 'Secondly, I don't think you can use a blanket approach. I think that . . . companies need to look at requests when there are circumstances that require individual attention, such as murders, paedophelia, abuse . . . . you need to look at each one of these individuals. If you're selling a product, which you are, a service and a product . . . you have that moral responsibility, if not legal responsibility. We need to give people opportunity to take action that actually would benefit society as a whole, individuals as a whole, next of kin.'

Moral responsibility. I think about all the times that an evolution of terms and conditions has been prompted not by a sense of moral responsibility, but by economic realities, by the threat of a user backlash due to a big story in the media, a story like Cambridge Analytica. I remember one of the events that happened just before Facebook shifted its policies on profile memorialisation, the announcement of which was made in the week of Hollie's murder, February of 2014. A father called John Berlin had been repeatedly unsuccessful in petitioning Facebook to create one of its special 'Look Back' videos for his son Jesse, a 21-year-old guitarist from Missouri who died in his sleep of unknown causes. Facebook's policy had been that they didn't create Look Back videos for memorialised accounts, so they initially refused. In desperation, John posted a raw, emotional plea on YouTube that went viral with millions of views.[182] That video pushed Facebook to change its previous stance, and they announced they would henceforth happily provide Look Back videos to the bereaved. Within the same announcement, they made the change that, serendipitously, allowed Nick to see Hollie's posts, even though his own Facebook account was only set up after her death.

'Starting today,' the announcement said, 'we will maintain the visibility of a person's content as-is. This will allow people to see memorialized profiles in a manner consistent with the deceased person's expectations of privacy. We are respecting the choices a person made in life while giving their extended community of family and friends ongoing visibility to the same content they could always see.'[183] The idealistic view is that Facebook was making these changes out of a sense of moral responsibility to its users and society. The more cynical take is that it's out of economic considerations – leaving memorialised profiles as they were might mean fewer special requests for access to data, less of a drain on the Community Operations Teams. Maybe you think the answer is obvious, but I harbour considerable uncertainty about not just the motivations of corporations, but about the fate of my own data under their oversight.

Ω

I'm not alone in my insecurity. Many people are so unclear that they're afraid to even avail themselves of features like memorialisation of a Facebook profile – whether they're setting up their own arrangements or poised on the cusp of letting a data controller know that someone has died. They're ignorant about the process or about the outcome of such a move, uncertain and fretful about what such changes will really mean. People fail to inform Facebook, leaving the profile entirely unchanged, as though the person were still alive, or managed by someone else, a friend or family member either already logged in or in possession of the password. In an effort to keep the profile 'alive' and to stay connected to their dead loved one, a bereaved individual may essentially end up impersonating them.

The number of people in that latter category leads us to delve yet further into a question that we've already touched upon, but which deserves closer examination. Traditionally, legally, the dead don't have the right to privacy. Has that changed?

Chapter Four
Behind the Portcullis

The producer was putting together a programme about digital legacy, and she was getting in touch because she hoped that I could clue her in about a few things. She didn't know much about social media and didn't have a Facebook account herself, so she was having difficulty interpreting some of the things that her source, a bereaved mother called Rachel, was telling her.[184] Rachel had lost her daughter Katie a few years prior and was distressed because something had changed about her level of access to Katie's Facebook account. 'Can I send you a file for background?' the producer asked. She pinged it over to me. I listened to five minutes of a now-familiar tale: a problem associated with a digital legacy, a confused family, a sense of helplessness that only deepened with every attempt at resolution. I rang the producer back to see what she needed. She was trying to understand how memorialisation would have changed Katie's Facebook account. What, exactly, could Rachel not see any longer?

I replied that if her mother had been on Katie's list of friends, which apparently she was, the profile should look just the same as it always had. Clearly, though, Rachel felt that something *had* changed considerably, and in a way that felt significant to her. Finally, I realised that the producer was using the phrase 'Facebook page' to mean both the material visible to everyone on her friends list *and* Katie's private Facebook messages. Was Rachel upset that she couldn't see Katie's Facebook *messages* any more? The producer wasn't sure. I wasn't sure. Rachel talked about Messenger as being a conduit to Katie's friends, and although there was nothing to suggest that Rachel was writing to Katie's friends from within the account, I thought with disquiet about Vanessa Nicolson, the woman who'd written about sending a message to her late daughter's boyfriend when logged in as her daughter, whose story is described in Chapter Two. I also thought about the research demonstrating that people often found it unsettling when a deceased person's social media continued being 'managed' by someone else. I had some compassion for bereaved parents who did this, ignorant or worried about memorialising the profile and not fully realising the potential impact on other people. Wanting to understand more about what was going on, I asked the producer to put me in touch, and eventually Rachel and I met at her hotel in London when she was in town.

Rachel's story was a mirror image of the Gazzards' experience in many of its facets. Katie and Hollie were about the same age when they died. Just like Hollie's profile, Katie's Facebook page contained images that were distressing to the family. One image in particular rankled with both Rachel and Katie's brother. Unfortunately, it was quite a dominant one: not the square-framed profile picture itself, but the larger, tone-setting banner photograph that spreads across the top of the page. Rachel and I pulled our chairs together to look at it. There was Katie on the left, grinning broadly into the camera but looking unwell in comparison to her usual appearance, her mother said. The backdrop of the photo was her room in the nurses' residence, recognisable by the décor. Hearts were Katie's 'thing', her mother said. On the right was a fellow nurse – whom I'll call Melissa – also smiling, arranging her fingers into a shape that echoed the hearts in the background. The room was where Katie died, shortly after the photo was taken, having injected herself with a carefully calibrated overdose of insulin obtained from the store cupboard of the hospital where she worked. Melissa was the one who'd given the family the first inkling that something was terribly wrong, leaving a garbled, hysterical message on Rachel's voicemail. Since Katie's death, Melissa had broken off all contact with the family, apparently even blocking Rachel on Facebook.

A troubling photograph that disrupted the family's ability to visit their loved one's profile comfortably was one of the most significant commonalities between these two situations, but there was something further. Just as with Hollie's profile, Katie's profile had been memorialised without an explicit request from the family. Katie's family had contacted Facebook to request removal or change of the banner photograph – just as the Gazzards had got in touch with their selective editing request – but to no avail and, to Rachel's recollection, they had never received a response. To this day, Rachel has no idea how or why memorialisation happened; she only knows that, once it occurred, she didn't have the access or control she once had and could no longer change the profile picture that so disturbed her. To add insult to injury, she hadn't even fully realised that she once did have the power to change it. Now that it was too late, I felt concerned about even mentioning this to Rachel, for fear it would be too frustrating and only add to her pain.

For all the similarities in the two stories, the priorities of Hollie's dad diverged from the needs of Katie's mum in one key respect. In my conversations with Nick, he never mentioned desiring access to private messages, the contents of Hollie's email accounts or her correspondence on Facebook Messenger. Lack of access to that sort of material was never something he cited as a problematic aspect of the family's wrangles with Facebook; he was only concerned with what was, and always had been, publicly visible. But Rachel was devastated when access to the messages was taken from her. On the recording the producer had sent me, Rachel had used a particular term to describe what the memorialisation of Katie's Facebook profile felt like. *Portcullis.* I didn't recall ever having heard the word before, so I grabbed my phone to look it up. 'A strong, heavy grating that can be lowered down grooves on each side of a gateway to block it,' said the definition.[185] I pictured the lines of corrugated metal barriers that are pulled down at night with great rattling and banging over nearly every shop window in my East London neighbourhood. They're secured with serious-looking padlocks, and they obscure anything that lies beyond.

'Portcullis' must have felt a particularly apt analogy to Rachel, because she employed the word again when we spoke in person. It was when we were talking about the caretaker role she'd assumed in looking after her daughter's personal data. As Katie's mother, she said, having ongoing access to the correspondence in Katie's various online accounts was only right and proper. One of the first disruptions to that access happened when she found herself locked out of Katie's Gmail account. Rachel believed it was because her ex-husband had changed the password, a claim I wasn't in a position to verify, but she was convinced that this must have been the explanation. Even when families are friendly and intact, the management of digital legacies can be a contested issue, but the nature of Rachel's relationship with her ex made things all the more difficult. 'I could have done that. Blocked him. I wasn't going to do that, but he did,' Rachel said. In her voice I heard the acrimony of the couple's parting, intermingling with the pain of their daughter's loss. 'Which meant I now no longer have access to her emails, which is useful for all sorts of things for me. I would want access to them. I would want to feel that I was custodian of her information.'

She paused. Was she picking up on something in my face or body language? Was I conveying something that made her concerned about what I thought? In fact, I was wondering whether this was an answer to what I'd been musing about when I spoke to the producer of the programme: was seeing *private* messages what Rachel wanted, what she was heartbroken about losing? She suddenly seemed keen to reassure me. 'I'm not going to go reading them,' she said. 'But I could have access to her *friends* through the Gmail.' She had spoken about that on the recording as well, the loss of a conduit, a channel through which she could continue to connect with the people to whom Katie had been close.

The loss of Gmail access seemed relatively insignificant, however, compared to how Rachel felt when she went online one day and saw a word on Katie's Facebook profile page that she didn't expect to see: 'Remembering'. The addition of this new word wasn't the only thing that had altered, and she quickly discovered that she couldn't see

Katie's private Facebook messages any more. 'It felt like a portcullis had clanged shut, like bang, you may not enter. That's it. All you can do, it says to me, all you can do now is look at this ghastly photograph, if you're in Facebook . . . . access denied. As a mum, I could only view her account and use it as anyone else would. For me, as her mum, I felt almost doubly bereaved, because somebody faceless had decided that this would happen now, and it has, and that's it. And I had no recourse. I had questions, but no one to ask. I didn't know what to do.' I hear the repetition in her account. As a mum. As a mum.

But it wasn't just the fact that she was Katie's mum that made Rachel feel she should have a special level of access to the account. She believed it was her daughter's wish. Her daughter didn't have a digital will, but her actions, Rachel explained, made it plain as day. Before Katie died, she took the password off her computer, logged into all of her email and social media accounts, and wrote a suicide letter in a document on her desktop. 'She was thinking ahead . . . . she was a very clever girl, and she clearly thought that through. She wanted us to have access to her Mac,' Rachel said. Once the laptop was returned to them by the police, Katie's brother used the information in her contacts lists to ring each of her friends individually, not wanting to convey the news on social media. 'I suppose that was the first instance of her Facebook account being operated by us,' said Rachel.

That phase of things soon drew to a close, however – eventually, everyone had been informed, the funeral was over and done, the immediate condolences and expressions of grief had been conveyed. The laptop and the accounts that Rachel accessed with it clearly became about much more than notifying friends about Katie's death. Rachel primarily referred to the Facebook profile as being a connecting thread to Katie's community of friends, whether through the visible wall or the back-channel messages, or both. While she didn't speak about reading her daughter's Facebook messages, and at one point in our discussion suggested that she hadn't done so, there were certainly other things on the laptop that gave her grieving mother insight into the pain of Katie's inner world in the days, weeks and months leading up to her death. For some of these aspects, Katie would not have needed to clear the path of passwords, and there are some things that she may never have intended her family to see.

It's unclear, for example, whether Katie made a deliberate decision to preserve her browsing history or thought about the contents of her iPhoto. In the heartbreaking list of pages she'd visited before her death, her mother could discern not only the depth of the depression that Katie had kept secret, and the certainty of her decision to die, but also her concern with achieving that aim as painlessly as possible. This window into her thoughts was poignant in the extreme to read. 'The pages that she had visited in the previous few days [were] very distressing, because it was to do with suicide methods and things, and dosages . . . . the physiological chances, where to inject into a muscle, that kind of thing,' said Rachel. 'I mean, she was so careful to plan it right. She gave herself the same degree of care and consideration at the end that she would a patient.' And the browsing history wasn't the only thing on the computer that provided Rachel with an insight into the tumult of her daughter's emotional world. 'She had photos on her computer. It was . . . very painful. I didn't know what I'd find on there. Tens of thousands of photos of herself, selfies, of herself looking at herself in the mirror . . . looking sad, the next one crying, looking at herself – it was heartrending.' Rachel caught her breath, sighed raggedly.

John Troyer at the Centre for Death and Society has another question that he asks his students. One of them was mentioned in the last chapter – what would be the one bit of digital material you'd leave behind, if you could only choose the one? Here's another: if you were to die tomorrow, and your parents were able to see everything on your laptop, everything in your password-protected accounts, how would you feel about that? John doesn't formally collect data on his students' responses, he told me, but he disclosed that the general trend favoured ambivalence rather than an embracing of the idea. I'm unsurprised; for whatever reason, I harbour strong feelings about family or friends being able to access and read deceased individuals' message histories, and I don't fancy my context-collapsed digital archive being available to my own partner or next of kin. Perhaps it's because of what Edina observed about physical letters as compared to digital correspondence. It depends a lot on your individual context, perhaps, but Edina and I agreed that our digital correspondence histories feel far more intensely personal, more comprehensive, more exposing than anything we have hanging around on paper. Interestingly, when Rachel tells me a story about a different legacy, my felt response is somehow less rigid, more forgiving. 'My ex-husband just gave me a box of [Katie's] belongings, this week . . . [saying] it's just rubbish in there, I don't want it. Well. Rubbish?!' Rachel said, her tone incredulous. 'When I actually had a look, it was photo albums. It was precious things. It was little notes and diaries and cards from her friends. It was wonderful stuff that she thought was important enough to keep, in a box, with a lock.'

With resistance and loathing but also with curiosity, I allow myself to imagine what it would be like to lose my own daughter at the age of twenty-one. I picture a laptop – or whatever device she will use for her data in the year 2031 – alongside the childish, padlock-secured diary she writes in now, both sitting on a table before me. Which would I feel comfortable opening – and why? In 1923, after years of frustration and failure, the archaeologist Howard Carter breached the seal of King Tutankhamun's tomb, chipping away at the corner of a doorway that had been shut tight against intruders for millennia. As he held a candle to the opening, the dim beam of light fell on to the perfectly preserved objects that lay beyond. The financier of the expedition, Carter's employer Lord Carnarvon, was hovering nearby and asked if he could see anything. 'Yes,' breathed Carter. 'Wonderful things.'[186]

Ω

What do we mean by 'privacy', a concept so sacred that in 1948 the United Nations enshrined it within the Universal Declaration of Human Rights? 'No one shall be subjected to arbitrary interference with his privacy, family, home or correspondence, nor to attacks upon his honour and reputation', reads Article 12 of that document.

'Everyone has the right to the protection of the law against such interference or attacks.'[187] The UN's declaration set a basic global standard for all nations and peoples, and indeed pretty much everyone seems to agree that privacy of some stripe is important to the well-being and autonomy of every individual. It's a cultural universal, found in some form in every society that's ever been systematically studied,[188] but for all its ubiquity, it shape-shifts from culture to culture, context to context, proving itself tricky to define and just as tricky to legislate. Part of its protean nature is related to its multiple dimensions: territorial privacy, referring to the spaces that help safeguard individual privacy; bodily privacy, which protects the dignity of the physical human person; and informational privacy, which asserts ownership and control of our personal data in service of preserving our dignity and integrity as individuals.

Hold on a second, though. Is all of this hopelessly outdated? Might you recall hearing somewhere that privacy is over, old hat, *so* last decade? Does privacy – particularly informational privacy – even exist any more in a hyper-networked online environment, and if we give it up so easily, doesn't that illustrate that we don't care about it any more? Sure, privacy was considered sacrosanct once, but in an unprecedented technological era when everyone is constantly tracking and keeping everyone else under surveillance,[189] has that particular sacred cow been slaughtered? In 2010, the CEO of Facebook – perhaps not the most unbiased pundit – made a statement that most news outlets interpreted as 'privacy is dead'. In an on-stage conversation with TechCrunch, Mark Zuckerberg said, 'People have really gotten comfortable not only sharing more information and different kinds, but more openly and with more people. That social norm is just something that has evolved over time.' He went on to question whether Facebook should bother with such outdated notions. 'We view it as our role in the system to constantly be innovating and be updating what our system is to reflect what the current social norms are,'[190] he said, leading the reader to wonder whether Facebook had 'innovated' privacy right out of fashion, or whether it was merely responding appropriately to its alleged demise.

He remarked that if he could do it all over again, he'd make Facebook public from the start, such was society's current lack of interest in differentiating public and private information. 'I think Facebook is just saying that because that's what it wants to be true,' said one commentator in the *New York Times*.[191]

On one hand, if you look at what people actually *do* online, you might be inclined to agree with Zuckerberg. In the quest for immediate gratification, people do indeed hand over their personal information for relatively small rewards, even if they've been burned in the past. When we're asked to express it in pure economic terms, we may not value our informational privacy particularly highly. In one 2013 research study, for example, it was found that Internet users in Spain reckoned the value of their browsing history as approximately equivalent to a Big Mac meal in a Madrid McDonald's – about seven euros.[192] Part of this low valuing might be about ignorance of the implications – if the participants in that study had watched *I Love Alaska*, or thought more deeply about the real-life consequences of their nearest and dearest having full access to their entire browsing history, their estimations might have been rather different. Irrespective of the price we place on privacy when we're asked about it, it definitely seems to matter to us when ructions at the boundaries of private and public occur, when we lose control of our information and discover that the personal data we've chosen to share has not been managed as we anticipated. When we disclose private information to a friend or trusted associate, we place our faith in the recipient to manage that information the way we want. If a friend has ever let one of your cats out of the bag in an unauthorised fashion, if you've ever spluttered, 'That wasn't *her* news to tell', you know that your right to control your personal information feels inalienable. And when the social contract around your personal data is between you and a company, you may be surprised and outraged when that organisation's information-handling practices violate your expectations.

Many Facebook users were incandescent when they discovered that data they thought was 'secure' had been used by third parties – or, in other words, when they discovered that they didn't actually understand the terms and conditions to which they'd agreed. It's not known how many users deleted or deactivated their Facebook accounts directly as a consequence of the Cambridge Analytica revelations, but research indicates that people who abandon social networks do so mainly because of concerns about privacy.[193] Many of those who *didn't* delete or deactivate definitely considered it for a hot minute, if anecdotal information from my own social networks is any valid indication.

The doubters' ultimate decision to stay is a demonstration of the privacy paradox,[194] the fact that privacy *concerns* don't necessarily predict privacy *behaviour.* People may care about privacy a lot and fear intrusion, but still grant access to their personal information. The privacy paradox means that anxieties about actual or possible privacy violations may push us to deactivate, but the concrete, immediate payoffs of continued participation may pull us back from the brink. It also means that Mark Zuckerberg may not have that much to worry about at the end of the day, but nevertheless in the wake of the Cambridge Analytica dispute, he embarked on an extensive 'apology tour', having rather changed his tune on the importance of users' rights to control their personal information.[195]

If the country of which you're a citizen accepts the Universal Declaration of Human Rights, and if you're a human, you'll have a legal right to privacy. Even if you don't live in such a place, I'd argue that morally, you have a claim on it. Psychologically, you probably require it. Take territorial privacy: even nonhuman creatures like rodents struggle and go into decline if they can't achieve physical separation from others when they need it,[196] and if you're a human animal who's unable to control your personal space, others' interactions with your body, or others' access to your personal information, there are significant psychological consequences.

Privacy is about setting boundaries for oneself, 'the state of possessing control over a realm of intimate decisions, which include decisions about intimate access, intimate information, and intimate actions.'[197] Infants and young children, developmentally immature and with only a nascent sense of self, don't have the wherewithal to make such decisions, but it doesn't take long to acquire both the ability and the inclination. At the age of seven, for example, increasingly aware of how I presented her life on social media, my daughter posted a sign on her bedroom door.[198]

I disregarded the instruction but didn't get away with my violation. She waited for her moment, confronting me when we were alone in the car.

'I know one of your secrets,' issued a severe-sounding voice from the car seat in the back.

'One of my secrets?' I said, blankly.

'You privately shared something that I said was private,' she said coldly. 'I saw that photo of the sign on my door. It was on Melanie's computer.'

'I'm sorry,' I replied weakly.

There was a pause.

'It's my choice,' she said, with emphasis.

One scholar has divided the concept of privacy into two types: resource privacy, and dignitary privacy.[199] Resource privacy holds that privacy is just a tool, with some kind of instrumental value: I will trade you a certain level of access to my private information in order to be able to use this service, for example. It's a Faustian bargain that occurs innumerable times per day in the digital world. That wasn't what was going on here. There was no transaction. There wasn't anything secret behind that door, nothing of utilitarian value that she was seeking to protect, preserve or trade. There wasn't anything sensitive about the sign, and it wasn't so original that I was stealing her intellectual property.

Instead, this exchange was about her assertion of the other kind of privacy: dignitary privacy. Dignitary privacy is a principle, the idea that determining one's own boundaries as one sees fit is *intrinsically* valuable. It was *her choice*. Fundamentally, her privacy in that moment had nothing to do with that sign and everything to do with self-determination. She had attempted to limit the audience with which it was shared, and I had overridden that attempt. Communications privacy management (CPM) theory has a name for that: boundary turbulence.[200] A violation had occurred that had shaken things up, and the ripple effect of that turbulence was that she didn't trust me for a while. We had a conversation to re-establish the rules, one consequence of which was that I sought her permission to use the photograph again in a blog post and in this book.

As my daughter eventually moves into adolescence, figuring out who she is and differentiating herself from her parents, I know that she's likely to share less with me than she does now, and she'll probably take even more umbrage at privacy intrusions.[201] That's just part of normal teenage development. When she becomes a legal adult, her conviction that she *owns* her territory, body and information will likely become ever more cemented. And then, as happens to us all, age or infirmity will start to chip away at her ability to maintain her boundaries, until death comes along and delivers the *coup de grâce*. As a child of the social media age, would she be likely to care if her parents could access her data if she were no longer here? While the phrase 'private life' is familiar, 'private death' is not really a thing. Is it?

Ω

A mere six weeks after the seal to the door of King Tut's tomb was broken, the Earl of Carnarvon was dead. Fuelled by the popular press, the public's imagination was captured by Arthur Conan Doyle's hypothesis about his sudden demise: Carnarvon had been struck down by the curse of the pharaohs, a consequence of disturbing the repose of the dead royal.[202] The curse of the mummy was apocryphal, of course. The Earl of Carnarvon actually expired from an infected mosquito bite, five years shy of the discovery of penicillin. Still, was there a problem with Egyptologists entering the tomb, disinterring the boy king, and transporting his body and earthly possessions around the globe? The colonialist insouciance with which certain powers whisked away ancient remains to distant museums certainly inspires embarrassed discomfort in modern times, but is there more to be concerned with here? Is it ridiculous to question how the mummies themselves might have felt about it all? In 2010, a medical ethicist and an anatomist got together to dissect the morals of just that matter,[203] and a number of popular news outlets picked up the thread, querying whether Egyptian mummies have a right to privacy.[204] Being a rather niche concern, and hardly a question that had been on everyone's lips, the rights of mummies didn't exactly evolve into a major news story or prompt changes in the law.

The idea that the dead might have a right to privacy, even those deceased whose associates and heirs have long since passed away as well, has never been taken particularly seriously, but it's more relevant now than it's ever been. Respect for the dead may be a moral principle that's generally valued in most societies, but formal privacy rights layered on top have traditionally been seen as unnecessary at least, over-egging it at worst. The extent to which times have changed, however, and the ways in which that change has happened, would seem to prompt a major reconsideration. The change should be fairly obvious: at no point in human history has there ever been such a volume of personal information gathered, stored, transmitted, and shared with or without our knowledge, on sites like Google, Facebook and countless other platforms.

Egyptian mummies left behind their linen-wrapped bodies and their bejewelled and gilt grave goods, but although surviving papyrus scrolls have taught us much about Egyptian society, the ancient rulers didn't bequeath us a whole lot of what we might term personal data in the modern sense. Slightly higher up on the personal scale, my grandparents left behind not only their worldly, physical goods, but a variety of autobiographically telling verbal and numerical records, including labels on scraps of fabric, a notebook with measurements and sketches for kitting out the interior of a VW Westphalia campervan, financial accounts and various other categories of paperwork, and that box of revealing letters from World War II. Maybe you would maintain that the mummy's artefacts, and my grandparents', should have been inviolate. Whatever your opinion on that, though, the intimacy of these materials pales in comparison to what's left behind when a digital-age human dies. The argument often put forward for why family *should* be able to access loved ones' digital materials is that it's no different to their getting hold of the deceased's physical, verbal and visual records, like the letters, diaries and photos Rachel had recently received back from her ex-husband. But is there something about digital records that makes them not just personal but somehow *hyper-personal* – and are they hyper-personal enough to inspire a wholesale rethink about the privacy of the dead?

Privacy may be a human right, but humans are mortal, living, breathing beings, sometimes referred to in law as natural persons. It's often assumed in law that only natural persons have legal personalities, and by extension only natural persons have full legal claims on privacy. Dead people aren't usually considered to have much, if any, interest in it. The cut-and-dried, well-established legal principles that few judges can or will question are the 'black letter law' that Edina referred to, and it's notable that black-letter laws often have hundreds of years of tradition behind them and hence may not take much account of the digital.

Take the territorial type of privacy, for example. If you're dead, you can't maintain control of your property, because there's no longer such a concept as 'your'. Furthermore, the physical space that you occupied in life is tangible stuff: the house to which you held keys, the land that you fenced off, the gate that you festooned with 'Private Property – No Trespassing' signs. After your death, all of this will pass on to your heirs, or the state, or revert to your landlord. Digital locations, lacking physical tangibility and perceived value, don't count as 'territory' in the same sense.

And what about privacy of your person, the right to be free from undue intrusions like degrading treatment, unwarranted searches, and physical assaults? Desecration of a corpse may be against the law in some jurisdictions, but your family or close associates, the parties who would actually experience the emotional impact of such an insult, are considered to be the victims of the crime. In my home state of Kentucky, for example, the test of whether criminal desecration has occurred is determined by whether it would outrage 'ordinary *family* sensibilities'.[205] The dead person themselves can presumably no longer perceive or be harmed by such slights to their dignity.

It might also surprise you to learn that, in many places, the dead themselves have so little right to bodily privacy that assaults upon that body may not count as a crime at all. In English and Welsh law for example, despite the fact that most people consider it a dreadful thing to behave disrespectfully towards the dead body, the legal system is 'incapable of dealing with cases of corpse desecration'.[206] These days, most of the time, actual dead bodies are rather difficult to access, and as such any abuses of them are, thankfully, rare occurrences. Digital remains, on the other hand, are far more accessible and vulnerable, as are digital memorials. When a twenty-five-year-old man from Reading posted videos and messages on YouTube and Facebook in 2011, mocking dead teenagers and their families, one of the places he targeted was the tribute page to Natasha MacBryde, set up by her brother. In the end, he was indeed convicted and sentenced to jail for eighteen weeks, but on a charge of 'malicious communication' rather than anything specifically to do with mocking the dead.[207]

Finally, there's informational privacy, referring to our ability to govern our personal data and to protect or disclose it as we wish. This used to be a far more straightforward matter, overlapping in some ways with territorial and bodily privacy. If you had verbal records of an intimate nature, you could keep them in a file cabinet on your property, or in some extension of your personal territory, such as a safety deposit box rented at your bank. You might keep personal photos locked away in a box, perhaps even keeping the key on your person, if those images were sensitive enough. Pre-digitally, only rarely might someone have asked you to give over personal information in exchange for goods and services, and then those someones were likely to be legal or other governmental entities. People could decide, moment to moment, whether to let privately telling information escape their physical lips in any given social interaction. Now, of course, the genie is out of the bottle. Our information has leapt the confines of our physical and territorial control, travelling independently, abroad in the land and impossible to completely summon home.

Like privacy itself, 'personal data' can assume different definitions and subcategories. One type is information that can be used to identify, locate or contact a particular person, sometimes referred to as personally identifiable information. Another category is sensitive personal information or sensitive personal data, which is information that you might or might not decide is someone else's business, data that is intimate enough that you'd like to choose whether you conceal or reveal it, to whom you disclose it, and under what circumstances. This is the 'maybe it's none of your business, and maybe I'll tell you, if I trust you' category, and under this umbrella the European Union's 2018 General Data Protection Regulation (GDPR) included your racial and ethnic origin, your political opinions, your religious or philosophical beliefs, trade union memberships, your genetic and biometric data, health data, and information concerning your sex life or sexual orientation. The GDPR, however, like most other data protection legislation and regulation, doesn't refer to anyone who's not a 'natural person'. That's for the EU member states to figure out. While the guidelines for how our personal data should be respected and protected when we're alive becomes ever more defined and regulated, what should happen to that data when we're dead is still largely up for grabs.

Does it matter, though? The usual argument put forth for not protecting deceased person's informational privacy is that there's no real
harm – essentially, the deceased won't or can't be hurt by such intrusions, because, well, they're dead. In the trolling case mentioned
above, it wasn't Natasha MacBryde that was the technical victim; it
was her family, the recipients of Sean Duffy's malicious communications. Following the no-real-harm line of thought, why would there
be any problem at all with having a free-for-all open-access policy to
the deceased's personal information? Is there any sensible argument
for why a dead person – not a 'person' at all, in law – should have the
'right . . . to preserve and control what becomes of his or her reputation, dignity, integrity, secrets or memory after death', otherwise
known as post-mortem privacy?[208]

Edina Harbinja reckons so, arguing that informational post-
mortem privacy is simply a logical extension of the control that the
dead have long enjoyed (at least the dead from countries that allow
considerable testamentary freedom): the ability to dispose of your
assets largely as you see fit, without having this dictated by the state.
'If property is an extension of an individual's personhood . . . personhood transcends death the same way . . . property does, through a
will,' she writes.[209] The same idea applies to copyright over intellectual
property, an area of law that is relatively well harmonised globally,
and certainly more consistent from region to region than the laws of
succession. As mentioned in the last chapter, after the person who
has created an artistic or literary expression dies, the rights to those
works passes to their heirs – sometimes for a particular period of
time, and sometimes for ever, as is the case in France. In that sense,
Edina claims, the 'legal person' *doesn't* actually cease to exist on the
point of death. 'Dignity, integrity and autonomy, do survive death,
sometimes even for an unlimited period . . . . Therefore, legal personality does extend beyond death and so should privacy.'[210]

When Edina and I spoke face to face, she referred to the relatively
hyper-personal nature of the digital data we leave behind. At that
point, I hadn't yet spoken to Rachel about the box of Katie's things
that she received from her ex-husband, but I still had a 'box of letters'

analogy in mind. Why should online material be considered any differently to physical objects that are left behind, boxes of correspondence like the intimate love letters between James and Elizabeth, my grandparents – items that next of kin could easily get hold of if they hadn't been destroyed? If no one writes physical letters any more and it's all online, why shouldn't the bereaved still be able to see those things, the types of things they'd always been able to see?

'In the online world, I argue that *autonomy* should be the first, most important value,' Edina said – the ability for the individual to decide. Certainly, taken together, the material on and accessible through Katie's laptop that was not already publicly visible, both the information that Katie deliberately and knowingly compiled, and the data that was incidentally stored, comprised a startlingly personal record, a window into an inner world that, in the last period in particular, was marked by vulnerability and suffering. 'There should be options for individuals to decide about those accounts,' continued Edina. 'But if there is not such a decision, then I argue that the next of kin should *not* have default access.'

So can we know, for certain, that Katie made such a decision about her post-mortem informational privacy? Her email was hosted by Google, but she had no Inactive Account Manager. On a 2017 survey conducted by the Digital Legacy Association in the UK, about 90 per cent of respondents hadn't done that either.[211] She had Facebook, but no Legacy Contact, and 85 per cent of the survey's respondents said that they'd made no plans whatsoever for any of their social media accounts. Katie didn't make a will, and there too she is in ample company – only 37 per cent of the Digital Legacy Survey respondents had written a traditional one, and only 2 per cent responded that they'd undertaken a 'social media will'.

On one hand, I find myself inclined to agree with Rachel's assertion that Katie made the decision that she wanted her information shared, but that's partly based on various biases and assumptions that I hold. Because I keep my own computer and accounts constantly locked down and share my passwords with no one, perhaps I find it difficult to relate to people who don't do this. It's easier for me to imagine that

Katie deliberately removed such barriers than it is for me to wrap my head around the idea that perhaps she didn't always keep those boundaries secure. But what if vigilance around protection of her personal data *wasn't* her usual practice? What if she always remained logged on when the laptop was in her possession, and what if her distress meant that she intended but forgot to log out? What if she was in too much turmoil to plan that aspect of things as carefully as her mother assumed that she had? Certainly, though, if she did deliberately remove those passwords and log into those accounts sometime in her last moments, this is a sequence of actions that one can readily imagine as meaning, 'I grant you access to this private material.' It seems entirely plausible.

Of course, things are not always so clear-cut. Around the time that Edina and I were speaking, there were stories in the news about a fifteen-year-old girl in Germany who'd committed suicide on railway tracks. Her parents had no access to her Facebook messages, but they were worried about her having been bullied before her death, and they wanted to see their daughter's electronic correspondence.[212] Facebook refused, and then the German courts did as well. In a commentary that privileges the emotional experience of the bereaved over the privacy of the deceased, perhaps particularly because the deceased was a child, Pete Billingham of the Death Goes Digital podcast and blog wrote, 'If you are a parent, you can understand and sympathise with the need to know if the answer is in those messages. It would eat away at you every day ... wondering if in the chat conversations you might find that sentence that could explain, guide or show you why. But if you do not know the password to your child's social media accounts, you are locked out of that world.'[213]

Eventually, in an important ruling handed down by Germany's Federal Court of Justice in July 2018, a judge agreed. Saying that there was no reason to treat the digital content on Facebook differently to physical letters, presiding judge Ulrich Herrmann ordered that Facebook should give the deceased teenager's parents access to the account.[214] It was precisely the kind of 'big fat lawsuit' Edina Harbinja had been craving, but even if this ruling makes the situation clearer in Germany, it may prove a controversial decision – and it remains to be seen whether other jurisdictions will follow suit.

Particularly when a death is difficult or unexpected, when there are troubling questions and no ready answers, many might argue that it's natural for a grieving parent to want to find out anything they can. What would you do? Hoping that answers will bestow elusive feelings of peace, after your world has been shattered, would you go looking for that peace in the exceedingly fertile informational territory of your child's online life? Whether you will ultimately find the relief you seek, or even more pain, is open to question. It's a gamble, whether or not you realise it, but would it be a gamble worth taking? And if you were prevented from seeking it by forces you did not anticipate and could not control, how would you feel? It's not difficult to imagine that you might feel the same way as Sharon.

Ω

Her daughter died four years ago, but Sharon's voice on the phone was still tight with anger, her throat constricted with the build-up of four years' worth of resentment towards Facebook. As we spoke, although we'd never met, I could picture her face, for in 2015 she and I shared space on a news spot about digital legacy, on BBC TV.[215] Sharon's daughter Amy was twenty-three when she died, the only child of a widowed mother, and although the circumstances of her death had not been violent and troubling in the same ways as Hollie's murder and Katie's suicide, nor had they been simple for Sharon to understand and accept. Amy was a formerly perfectly healthy twenty-three-year-old woman, who was carried off by a rare heart condition in a matter of days, leaving her mother in shock.

Sharon could barely process how such a thing could happen. What had Amy been doing in the days before her death? How had she been feeling? Was there anything, anything at all, that could cast light on what had occurred? Amy's Facebook account was not yet memorialised, so if there was anything in her Messenger that could give Sharon some more information, some answers, maybe some comfort, that could potentially be available to her – with the login details. Sharon didn't have those, but she made an assumption that many parents might share: if she wrote to the social networking site, surely they would give them to her, as her daughter's next of kin. She sent me a copy of the report she filled out at that time.[216]

Other Notification Complaint: #5834023 . . . .[217]

What would you like to report? *Other*

What right is being violated or infringed: *Please help me urgent my 23 year old daughter has died and I need her log in passwords*

Check if applicable: *no access*

Provide information sufficient for us to locate the content on Facebook: *please help me urgently*

How does the content violate or infringe your rights?: *please help me urgently*

Pick one: *rights-holder*

Please help me urgently. Please help me urgently. The pathos in Sharon's repeated phrase, the appeal for help filled out on what ultimately proved to be the wrong form, the assumption that she would have the power to get what she felt she needed – erroneously identifying herself as 'rights-holder' – feels devastating. When she gives me more context, it's even more poignant to realise how soon after Amy's death this desperate appeal for assistance happened, for the funeral had not even occurred.

On one hand it's encouraging that – unlike some other parents with whom I have spoken – Sharon does recall having received a response. On the other hand, the canned-sounding message is a bit of an additional knife to the heart, like hearing a desperate call to emergency services being answered with an 'all lines are busy now' recording. Whether the reply was generated by a human being or a chatbot, it's hard to say, but I've since been assured that chatbots were not involved at Facebook in these types of communications at that time.[218] The email starts as one might expect, saying that Facebook is very sorry to hear about Sharon's loss, but then it quickly moves on to emphasising Amy's right to privacy.

The choice of the word 'right' is interesting, given that this right is conferred by Facebook's own terms and conditions, rather than being enshrined in law (where the dead in many jurisdictions are concerned), but the message didn't refer to this. Facebook takes protection of its users' privacy extremely seriously, it said, seeming to imply that it was yet to be proven that the user whose information was being requested was actually dead. If the user was indeed dead, Facebook could only provide information about the account content to 'authorised representatives' of the estate. Even for such authorised persons, the email went on, simply making requests or completing documentation would not result in immediate success – obtaining content from a deceased person's account would involve a lengthy process and a court order. The email signed off with an individual's name – we'll call her 'Jackie' – identified as working for Facebook's User Operations department.[219]

Undeterred, Sharon persisted. She wrote back to Jackie, thanking her for the swift response and explaining that the funeral was only days away. 'I really do need to get urgent access to check her history, etc . . . This is certainly no hoax,' she assured her,[220] clearly under the impression that as long as she could demonstrate that her daughter was actually deceased, Facebook would recognise her need and assumed right as a mother to access her information. She suggested that the relevant department check Amy's Facebook page as confirmation that her daughter had indeed passed away – there was a link there, Sharon explained, to a headline in the local newspaper about her untimely death. Jackie wrote back to say that with the information Sharon had provided, she couldn't help with Sharon's request.

Despite some further back and forth, Sharon never did get the login details to Amy's account. All that her correspondence with User Operations accomplished, in the end, was to alert Facebook that Amy had died; per its policy, the site then proceeded to memorialise the account – a consequence that she did not intend. Sharon was also sent a link where she could request removal of a deceased user's account – like Hollie Gazzard's family, this was the opposite of what Amy's grieving mother wanted. Four years later, Sharon said that she still looked at the profile frequently, describing it as 'vital – I'd be lost without it'. Ultimately, Sharon got nothing she asked for, topped off with a couple of things she didn't: a signpost to a link where she could cut her lifeline to her daughter, and profile memorialisation, which means she is no longer sent reminder messages on Amy's birthday. Sharon feels that she's not the only one who's affected by that. 'It was her birthday ten days ago,' Sharon said to me. 'Facebook didn't say, it's Amy's birthday today. That no longer happens. Really, she's still on Facebook, and it was still her birthday.'

Sharon was aghast at Facebook's refusal to give her access to her daughter's messages. Part of that seemed to be because she could not comprehend how or why they would consider the deceased's privacy over and above the emotions and wishes of the family. 'It annoys me, that I wasn't allowed to look at the private messages. They say it's data protection, protecting that individual's memory.... You see, it's wrong. It's *wrong*,' Sharon said. Her voice was emphatic, even as it trembled slightly. 'They should respect the wishes of next of kin. It's *wrong* that [grieving families] should have to do this. What Facebook will have done is just got legal bods behind it. We have to have a *procedure* when someone dies.'

Didn't they *have* a procedure? I asked, tentatively. I pointed out that, looking at the email thread, they seemed to have had rather a lot of procedure, as imperfect or as impersonal as it might have been.

'They have a legal procedure,' Sharon acknowledged. 'But they don't have a compassionate one. When I lost her, it was by far the worst thing. Because we were so close – we were really, really friends. She respected me as her mum, very much so ... she absolutely respected ... ' Sharon paused. 'I was heartbroken,' she said.

I flashed back to a clip from the news story in 2015, to an image of Sharon's finger pointing at a screen, showing the gap in the list of Sharon's own Facebook friends. After her death, Amy inexplicably disappeared from that list – the last in a string of insults and injuries that Sharon felt the social networking site had inflicted on her. She told me how she appealed to Facebook for help once again. Once again, she recalls there having been a failure to communicate, no apparent hope of resolution. It was only once Sharon took her story to the BBC that Amy was restored to Sharon's list of friends. I thought back to Rachel's repeated assertion 'as her mum', mentally aligning it with the term Sharon selected to describe herself on her initial inquiry to Facebook: 'rights-holder'. I thought of Sharon assuring me how much Amy had respected her, how close they were, and I wonder if there was another layer of meaning there: Amy would have wanted me to have access to those messages. She would have wanted that. She trusted me. I'm her mum.

Commiserating with Sharon's suffering, hearing her pain, I felt like a double agent. There's a twinge of guilt – it was there then, when I was speaking to her, and it's here with me as I write, now. In one sense, of course, I'm on Sharon's side. I'm a mother of an only child, a much-beloved daughter. I'm a psychologist, a practitioner who works with bereaved people. I empathise deeply and effortlessly with her entirely normal desire to access those messages, to try and resolve the painful doubts and questions in her mind about her healthy daughter's rapid decline and death. I know enough about grieving in the digital age to understand the pain of actual – or anticipated – loss of access to a digital legacy. Although she was never shut out from viewing those parts of Amy's profile that she had always been able to see, when her daughter's name disappeared from her own list of friends, Sharon experienced a version of the traumatic 'second death' that is feared and experienced by so many mourners in the digital age. I agreed that the Facebook message thread she sent to me felt rather cool and impersonal, and I felt angry that it wasn't easier for bereaved people to locate and navigate information about what to do and how to contact Facebook after a loved one's death. I felt compassionate and curious about the psychological reasons driving Sharon's desire to see those messages.

Even with all of that, though, I wasn't sure that I believed in her right to access the private data that Amy had generated in life. Like Katie and the vast majority of the online population, Amy hadn't made a social media will or expressed her wishes on any particular platform. But unlike Katie, Amy hadn't logged on to all of her accounts before she died, and had never shared her passwords with Sharon, however much she respected and loved her, whatever the strength of their mother–daughter bond. In this instance, it's not as easy to argue that Amy made a decision about whether she wanted her private data to be accessed. Maybe you agree with the moral arguments for post-mortem privacy, and support the proposition that, in some future version of commonly adopted law, hyper-personal informational data should be treated as legal extensions of the person that stretch beyond their death, just as physical and creative property do. Maybe you don't agree, and so don't feel the same reservations that I do.

Part of me wanted to ask Sharon about this, and Rachel too. But depending on how you see it, I was either too cowardly a journalist or too sensitive a psychologist to broach that topic. I drew my conversations with both Rachel and Sharon to a close without ever daring to ask them the questions that kept percolating in my brain, the questions that I suspected would elicit anger, hurt, defensiveness, or all three. What if it had been the case that your daughters actually wouldn't have wanted you to read their messages – and what if it were the case that they *still* don't? I asked neither mother whether they believed in an afterlife, but Walter talks about how we perceive the angelic dead as watching over and protecting us, the living.[221] If Amy or Katie were looking down on their respective mothers now, and all of Rachel's and Sharon's efforts to gain access to their daughters' messages were still blocked, what would that mean to them?

In early 2018 I attended the Information Ethics Roundtable, an occasion for scholars of various stripes to come together and discuss big data, algorithms, surveillance, data protection and privacy. One of the colleagues on my own panel presented the results of a survey conducted in Israel, in which 478 Internet users were asked about their wishes for their own personal data after death. One question asked respondents to consider who they would want to have access to various platforms after they die. Who would want their parents, specifically, to see the content of their social networking sites after they were gone? Fewer than a quarter were happy with that idea – 22 per cent, compared with 68 per cent who said that they would feel comfortable with a spouse or partner seeing it. What about the messages contained in email or other private channels? Fewer than a fifth of respondents felt comfortable with that, coming in at 19 per cent, even though the per cent that would feel OK with a spouse seeing private messages held steady at 68 per cent. Apparently, the tendency to form a relatively thick informational boundary between ourselves and our parents, so apparent in adolescence and young adulthood, is something that we maintain throughout our lives, for their or our greater comfort, and those preferences extend beyond our demise.[222] And what was the most frequently cited reason for these survey participants wishing to deny others access to this information? You guessed it: a reason that – once you breathe your last – is often nonexistent in law, and that until recently made little sense at all where the dead are concerned: privacy.[223]

Ω

Here's the final complication, though, and although it's the last hurdle of complexity to be tackled here, it's been integral to this issue all along. The clues to what it is are in terms that have been used not just throughout this chapter, but throughout the book: 'messages', 'social', 'networking'. While a journal may be kept purely for oneself, a message is always *to* or *from* someone else. It's nonsensical to socialise with yourself, impossible to be a network of one. If the material visible on a deceased individual's Facebook Timeline is usually co-constructed amongst multiple individuals, the authorship shared amongst potentially hundreds of people, message histories always involve at least one other.

We're not just talking about a deceased person's informational privacy, about that individual's identifying and/or sensitive data. This involves the data of anyone and everyone with whom a deceased individual ever corresponded online. Remember the Ellsworth case from Chapter One, in which the courts ordered Yahoo! to turn over the downloaded contents of Justin's email accounts to his father? Part of the reason the provider may have been so resistant was because the majority of those contents were nothing to do with what Justin's father needed to organise his estate, instead consisting of correspondence with multiple other individuals, some of whom John Ellsworth would have never met. Some of these individuals would have also signed up to Yahoo!'s terms and conditions and would have had certain expectations about their correspondence being private. It's not likely that any of them ever thought about their emails finding their way into the hands of his family if something were to happen to Justin; not until, that is, he died, and the story of the court case hit the headlines.

It's not the case that people don't care about privacy in the digital age, even those from a younger generation who have been accustomed to social networking from an early stage. In fact, younger people have been shown to be *more* protective of their online privacy than their elders.[224] In the same year that Mark Zuckerberg was declaring that privacy was no longer a social norm, research was taking place in which 71 per cent of young adults reported that they had taken steps to restrict their privacy settings, and 47 per cent of blog users (defined as including platforms like Facebook) said that they had retrospectively engaged in 'blog scrubbing', deleting comments and posts that they'd made but about which they'd subsequently had second thoughts.[225] Aware of the potential social impact of disclosures, on ourselves and on others, we assess, reassess, calibrate and adjust what information about ourselves we put out there. Conscious of potential problems but driven by the benefits of participation, we constantly tot up both sides of the balance sheet: are the social benefits I'll get worth disclosure of that bit of my private information? And, crucially, whenever I choose disclosure, *to whom* am I disclosing? For that bit of data, who's in my chosen sharing circle?

Here's where communication privacy management (CPM) theory re-enters the scene, because once again it's relevant, and it's easy to spot why. CPM theory explains that when a person shares 'private' information, personal data about themselves, it fundamentally changes the nature of that information. The original owner no longer holds sole possession of it, and it will never again be 'private' in the same way. Once you tell someone a piece of information that was originally only yours, your chosen recipient or recipients become joint shareholders or stakeholders in that information; the privacy boundary has altered for ever. If the disclosure took place in a one-to-one email exchange you had, the privacy boundary has expanded to two people, a dyad. If it happens on a friends-only Facebook post where your friends list numbers 432, and if no one else is tagged to make that post visible more widely, the privacy boundary has swelled to 432.

Interestingly, you very rarely see anyone say something like, 'Please don't forward this email to anyone else' or 'Please do not screenshot or share this post with anyone else.' They don't generally need to. For those acculturated to such mediums of communication, it's implicit, and nearly everyone understands that they are responsible for upholding the privacy boundaries of the group. When someone breaks this tacit agreement and spreads the originator's personal data outside of the circle, it could be because of any one of a number of reasons. It could be because of ignorance about the context, unawareness of the implicit rules, or failure to negotiate boundaries when things are uncertain or fuzzy. Occasionally someone knowingly engages in a more flagrant, deliberate violation. In any case, people who break the circle of trust – the unspoken, social rules of data protection that aren't dealt with in a platform's T&Cs – are likely to get negative feedback or punishment, through word or deed. The original owner of the information may even readjust their own boundaries in a way that excludes the offender.

But when the originator of the data isn't alive any more to slap the wrists of boundary violators, and to redraw their own boundary lines, does the responsibility still lie with the joint stakeholders? Will they maintain the assumed, reciprocal pact to protect information that was shared within the group? I think of Melissa, the friend in Katie's banner photograph, a person that Katie seemed to trust. I remember Rachel talking about her efforts to communicate with Melissa after Katie's death, and I empathise with this grieving mother, reaching out to Katie's friend via Messenger, searching for the answers she so desperately needed. And I also recall Rachel telling me about how, eventually, Melissa blocked her on Facebook.

Of course, the associates within a deceased person's inner information circle might not just be worried about protecting the data of the person who's passed away. They might very well be concerned for their *own* privacy. In an analogue, paper-based, snail-mail time, families would have had easy access to photographs and any surviving diaries. Acquiring the letters their loved one had sent to others would have been a trickier matter. Instead, they might have suddenly found themselves in possession of the letters that myriad others wrote *to* their dead family member, leaving family to fill in the blanks. On one hand, such material might not have much emotional resonance for or impact on those left behind, not being authored by the person that they lost. Sometimes they might be comforting. Sometimes they could provoke more upset, prompting questions and doubts that might prove difficult to resolve in the dead person's absence. Who was this person? Why was he or she writing to my daughter, son, wife, husband, father, mother? What does that paragraph mean? What was their relationship? In any case, relative to today, there wasn't likely to be that much of it, and the one-sided nature of the correspondence left wide gaps in the narrative. Now, it's an entirely different story. We get both sides, conveniently stamped with time, date, and often even location. Hyper-personal, indeed, involving the personal data of many more people than just the one who's died.

Just as was probably the case for Justin Ellsworth's friends, when we write to our intimates online, we rarely consider the fact that our audience may expand significantly should something happen to our correspondent. Sometimes we might contemplate, when typing out an especially private message or particularly sensitive disclosure, what it might mean if someone else were to see it. Technologically mediated communication on a variety of devices can cause messages to fall into someone else's hands – and that unintended audience could easily identify us and contact us with any questions or grievances they might have. That might give us pause. But we hardly ever extend our considerations to what might happen once our correspondent *dies*.

Let's assume for a moment that a significant person to whom you write is a staunch protector of their informational privacy, armed to the teeth with layers of password protection, but that someone else with a persuasive claim on that data were to wish to see it after they died. If future laws were to be passed that confirmed that the dead have no ongoing right to keep their digitally stored personal information private, how might that change the freedom with which you communicate to your friends in life? What kind of a constraint on our personal autonomy and liberty might that constitute? On the other hand, if legislation is someday passed that enshrines post-mortem privacy in law, how many grieving families would find themselves with no access whatsoever to anything their loved one ever wrote, with no ability to see any recent photographs of them? And in an age when privacy of individuals so frequently involves numerous joint stakeholders, how can we begin to get our heads around the privacy of the dead?

Ω

If your head is spinning, then you can probably relate to the reluctance of the European Union to tackle post-mortem privacy in its latest data protection regulations, its eagerness to let member states fight their own way through this particularly thorny thicket. You can probably sympathise with the Law Commission in the United Kingdom, launching a public consultation in 2017 with the entirely laudable aim of modernising the law around wills, but losing its nerve when it came to the bit about digital assets, saying they weren't intending to regulate those for the time being. Edina Harbinja had a big eye-roll at that, highly unimpressed at the Law Commission's decision at what she sees as such a critical juncture, when change needs to happen sooner rather than later.

Having just negotiated the not-insubstantial challenge of design-ing some kind of coherent tour to take you through this territory, though, I have some sympathy for all of the deeply confused and con-flicted humans behind the faceless regulations. In an entirely new and endlessly complex environment, we're finding it virtually impossible to understand all the risks, benefits and trade-offs that affect our privacy in life, much less in death. When the boundaries set down by a living person are redrawn after they die, the turbulence can be heavy indeed, and the hyper-personal data left behind can be a source of great comfort, or great pain. One hundred years after the archaeologists entered King Tutankhamun's tomb, as we enter the valley of the data of the dead and push open the doors we find there, we too – like Lord Carnarvon – experience both the wonder, and the curse.

Chapter Five
Stewarding the Online Dead

For some time after her death – too long, in my mother's opinion – my grandmother Elizabeth lay in an unmarked grave, the only indication of her presence a rectangle of turned earth with newly sprouting grass. Maybe the ground needed to settle before a memorial stone could be laid, but my mother wonders whether her father, James, was simply unperturbed about such things. 'I got the idea that he didn't really care about a gravestone,' she said. 'He had never thought that much about the gravestones of his parents.' However indifferent he may have been, my mother wasn't happy with this state of affairs, and she let her dad know it. Together, the two of them set out to redress the lack.

By that point, James's original nonchalance seems to have shifted considerably, and they arrived at the monument company poised to resolve the situation to everyone's satisfaction. Given his lackadaisical starting point and his unpretentious nature, one might assume that James would want to get the job done with a minimum of fuss and expense. In the wake of his earlier hesitation, however, he now seemed to feel a social or emotional imperative to create something more significant. 'At that point,' my mother said, 'he didn't care how much it cost. He just wanted to make something that was really, really special, something that everyone would be pleased with.' Although she didn't say it explicitly, I suspected that some emotionally fraught conversations had taken place between father and daughter to get them to this point, ready to make a decision about how Elizabeth – and, in due course, James – should be commemorated in stone.

Indiana's regional limestone ranges in colour from grey to buff, and much of the state is blanketed with it – whether in natural or built environments, you can spot it everywhere. Bedford, Indiana, not far from where my grandparents lived, touts itself as the 'limestone capital of the world'. In the 1920s, the quarrymen's association based in that town produced an educational monograph that described Indiana limestone as the 'aristocrat of building materials'.[226] It sounds like a fitting selection for two people who had lived their whole lives in the southern part of that state, but my mother distinctly remembers the monument company discouraging them from using it for Elizabeth's headstone. 'It wears away or deteriorates in some way . . . it lasts 400 years instead of 1,500,' my mother said. 'They won't use it any more. Everything is marble these days, or granite; everything is going to last *for ever*.' I told her that the designers for our kitchen extension in London also used a longevity argument to talk us into marble or granite countertops, telling us that this would be our *forever* kitchen. With the scepticism of someone who has undertaken multiple kitchen refurbishments in her time, my mother reacted to this with hilarity. 'We're all trying to get permanence in some way, because many of us don't really believe in the afterlife these days. I mean, what I'd like to do is do my kitchen counters in granite. Then you can just engrave that for our headstones.'

My mother might crack jokes about permanence now, but at the time she and my grandfather were apparently sold on the hope of perpetuity, and so it was that they found themselves meeting with a proper artist, a man they would eventually commission to sculpt Elizabeth's memorial from marble.[227] Visiting him at work in his studio, my mother was startled at his appearance: covered from head to toe in marble dust, white as a ghost and oddly reflective, glowing in the light. There was a reason for the sculptor's eerie luminescence, for the particles covering his body were tiny flecks of Colorado Yule, a stone only found at a site high in the mountains of Colorado, able to be quarried only with huge difficulty and at great expense. Stunningly white, composed of 99.5 per cent calcite, Colorado Yule is chosen for important state and national landmarks in the United States when only the best will do. In the early 1900s, the architect for the iconic Lincoln Memorial in Washington, DC decided that the only fitting material to clad the exterior of this important edifice, to create a shining symbol of Lincoln's defence of democracy, was elegant, pure Colorado Yule. Given the cost of the material, he met considerable resistance, but having successfully argued that the aesthetic properties of the stone made the expenditure worthwhile, he won the day. By the time the project was complete in 1922, it had cost $3 million, equivalent to over $40 million today.[228] So my mother wasn't exaggerating when she said that my grandfather spared no expense.

Having settled upon this premium stone, it only remained for them to select the sculpture's form. Ultimately, James made the decision for the monument that would one day mark not just his wife's final resting place, but also his own: a swan. Swans, my grandfather said, mate for life. My mother recalled her first impression of the swan sculpture. 'Wow!' she remembered thinking. 'This is *really* white!' It was her first impression when she saw it in the studio, when the artist was sawing away at the surface to create the stylised wings folded tight against the swan's resting body, its neck arranged in an attitude of repose. She noticed its brilliance even more when it was settled into place, where it contrasted starkly with the surrounding rows of identical headstones with gently curving tops, like a brilliant white boat bobbing on a sea of grey waves.[229]

The fact that the swan was out of keeping with its surroundings wasn't a problem in and of itself. In some cemeteries the memorial architecture can get competitive – the status-symbol monuments of Père Lachaise in Paris come to mind – but Walnut Ridge Cemetery in Jeffersonville, Indiana isn't the sort of place where families mind too much about keeping up with the Joneses, and the residents of the plots don't care. The more significant issue didn't emerge until a while later, after my grandfather had also died. Over time the piece proved to be a ready host for all manner of matter. Grime and organic material lodged in and bloomed on its coarse-grained surface. 'At

first it wasn't too bad,' said my mother. 'It just looked kind of shaded, and it helped keep it from looking *too* white. The darker creases emphasised the graceful features. But then it just looked forlorn. Like it had been forgotten.'

It had never occurred to my mother that the swan would need to be maintained. The employees of the cemetery were just caretakers of the grounds, and they weren't used to being asked about more diligent grave maintenance. My mother called the artist and dutifully ordered the cleaning product he recommended, but with her inexperience and no water source nearby, her attempts to clean it herself failed. No doubt the responsibility of keeping the Lincoln Memorial tidy occupies a team of US federal employees, but my mother didn't have anything like that. All she had was herself, a bottle of cleaning solution, a couple of hands-off cemetery employees, and a grungy swan.

Luckily, a family friend had the know-how and offered to help, and the swan was restored to its original appearance. My mother doesn't know how the friend did it, and nor does she remember the name of the cleaning product, but at least the swan was pristine once more, sticking out like a sore thumb, the price to be paid for being well looked after. But who will do the job next time, and the time after that? We're a geographically scattered family, bringing up a generation of kids who never met their great-grandparents. As much as she appreciates the swan's beauty, my mother doesn't think she would make the same decision if she had to do it all over again. And Colorado Yule may have overpromised and under-delivered on the 'for ever' score as well. On her visits to the grave, my mother notices a fine dust scattered across the base of the sculpture, standing out against the polished black marble base. As it weathers, the swan is shedding grains of incredibly expensive white marble. 'Maybe perpetuity will find a much smaller swan,' Mom said.

I told her that in countries like Norway and Sweden and Germany, people's dedicated graves expire after twenty or twenty-five years, unless the family pays for this time to be extended. If a maintenance fee is not forthcoming, the bodies are turfed out and cremated, or placed into a mass grave, and the plot is reused for the more contemporary dead, the ones more vividly recalled by the living.[230] This can come as a shock to Americans who, by virtue of living in a massive country, still have the luxury of giving many generations of deceased citizens individual resting places. Smaller countries can't afford the space. Certainly Mom was initially shocked, having never heard of grave-recycling practices, but then she laughed. Having grappled with the swan problem for two decades, and having lived for eight, she has become more pragmatic about these things. Even so, the thought of the swan falling into neglect troubles my mother. When she and my grandfather made their decisions, neither knew that the monument would be high maintenance. Neither realised that when they commissioned it to be created, they were also creating responsibility, a job, an obligation. My mother feels that obligation keenly and is a willing steward for the swan, but twenty years from now, she's unlikely to be here. 'I don't spend a lot of time thinking about it, but after I'm gone this is probably not going to be taken care of, unless somebody else in the family decides to take that responsibility,' she said. 'As long as someone's here and someone *cares*, that can be done. After that, nobody's going to know who these people are.'

Years from now, in my own old age, I wonder if I'll still travel across the Atlantic to the place of my birth, the country where my grandparents lived and died. If I do, will I feel moved to visit the

Black Swan of Walnut Ridge, as it may then be known? Gazing at it, its natural luminescence obscured by decades of grime, will I feel a twinge of guilt? Will I have an uncomfortable suspicion that I've evaded a responsibility that should have been mine?

Ω

In the wake of my conversation with my mother, some of her worry about the swan transferred to me, the eldest and geographically furthest distant grandchild of James and Elizabeth. Nevertheless, the upkeep of my grandparents' monument is a challenge that, if I choose to accept it, should be relatively straightforward to meet. It's physical, tangible, and in a fixed location; it's a structurally uncomplicated object that's relatively hardy despite the shower of fine dust that drifts about its base. A little bit of money and a set of instructions to a nominated caretaker should suffice, even if I still live thousands of miles away.

A non-degrading virtual swan in an online cemetery that I could maintain and visit remotely would have been a damn sight easier, as bizarre a concept as that would have been to my grandparents. Whatever their differences, though, what do American cemeteries, the online World Wide Cemetery, Facebook, legacy websites, and the Jeffersonville Monument Company – the now-defunct business that upsold my mother and grandfather a Colorado Yule marble swan – have in common? They have all suggested, and some even promised, that something of us can endure for ever. But if we stick around in anything like perpetuity, who is responsible for looking after us? Both offline and on, the dead and their legacies need their stewards.

Those who care for the dead are bound to do so through many different ties. The first caretakers are those who look after someone's physical remains during their last journey on earth: family members and death workers of various stripes, their roles clearly defined by tradition, convention and law. After disposition of the body, care and attention turn to the estate, and here too people tend to know their place, understand their fiduciary responsibilities: executors, personal representatives, next of kin, administrators. The cemeteries where grave plots or columbarium niches have been purchased are tasked with protecting and maintaining memorials for as long as they've been contracted to do so. Beyond this point, though, we enter a murkier territory, particularly where the online realm is concerned. Who is responsible for safeguarding the more intangible, sentimental, *digital* elements left in the wake of a person's life, the representation or memory of that person in the world? Who owns, curates or tends digital remains? Who guards the gate, grants or denies access, prevents desecration, preserves and honours the memory?

Light on clarity, but potentially heavy on responsibility, this kind of stewardship over digital remains and memorials may prove far more significant and consequential than the maintenance of any physical burial space, no matter how elaborate the monument or how expensive the stone. Stewards may be bound to care for the dead through kinship, friendship, professional responsibility or the terms of a contract, but let's start with family. I'm sure there are days that Susan wishes that the only things she had to deal with were a marble swan and a few ring binders full of paper letters, for her stewardship responsibilities are far broader than my mother's, in more ways than one.

Ω

Susan got in touch with me when I put a call out on Twitter, seeking any experiences that people might wish to share about mortality in a digital context. Her recent history of loss encompassed a cascade of untimely deaths in quick succession. The grim reaper arrives in many guises, but when he visits Susan's nearest and dearest, he's most often cloaked in cancer. Melanoma took her son-in-law, the husband of her youngest daughter Jonnie Sue, in 2004. Then came the second knock – Jonnie Sue was diagnosed with cervical cancer, stage IV, a formidable challenge for even the specialised cancer treatment centres to which Susan travelled with her daughter. The stress was significant for not just Susan, who kept her day job while she was helping her daughter fight for her life, but also for Susan's husband John, looking after two teenaged grandsons who'd already lost their father.

Given the strain he was under, Susan thought, it was understandable that John was losing weight. By the time his colon cancer was diagnosed, tumours had spread to his lungs, liver and brain. Each struggling with their own advanced illnesses, and each conscious of the suffering of the other, the eventual deaths of father and daughter were devastating in their concurrence. Although he was diagnosed second, Susan's husband died first. A mere fifty-eight days later, Jonnie Sue – a daughter named after both of her parents – died in Susan's arms. Even after all this, fate seems to have made up her mind to be cruel. Susan's best friend Emily, a local teacher beloved by everyone and Susan's constant pillar of strength, was carried off by brain cancer a year after Jonnie Sue died. Son-in-law, husband, daughter, best friend – all within a span of eight years.

Trying to find ways of dealing with such an enormity of loss, Susan started writing a blog about grief and coping with death.[231] That was certainly one way in which death met digital for her, but it wasn't the main reason she'd responded to my tweet. 'So much of this loss,' she wrote in an email to me, 'played out on social media, and continues to do so.'

That wasn't so much the case at the beginning. When cancer initially launched its salvo upon her family, social media sites like Friendster and MySpace were poised somewhere between infancy and falling out of fashion, and Facebook hadn't been launched to the public. Susan's son-in-law died without any significant digital footprint to repurpose, so the family constructed a memorial website for him, keen to provide a support for mourners and an online legacy for his two young sons. At the opposite end of the time spectrum, at the point I spoke with Susan, she wasn't actually too involved with social media either. It was just after Christmas, and Susan had recently temporarily deactivated her Facebook account. To grasp why, it's necessary to understand what happened in the middle, during which time Susan not only joined Facebook, but came to experience it as a constant presence and even pressure in her life, something she spent more time and effort on than she probably ever dreamed she would.

At first blush, this may seem a bit odd. Here was someone with no pre-existing social media presence, a stranger to Facebook and indifferent to Twitter. Although not averse to technology, social media wasn't Susan's thing, and after her daughter became ill, she hardly had much time on her hands, balancing work and care. Under those circumstances, how did Susan find herself being responsible for not only a profile page of her own, but multiple support pages? The answer lies in a word that she used in our interview even more than she used the word 'grief', a noun she employed more frequently than any other: 'community'. Long before there was any such thing as online social networking, Susan and her family seem to have been the beating heart of their Ohio community. Her high school boyfriend, the man who would become her husband, was an American football star, holder of multiple state records. When John and Susan married, they stayed in the area, and their three children went through the local school system, including Jonnie Sue. She was the 'darling' of high school, Susan said, the homecoming queen. Everybody knew their family. So when Jonnie Sue became ill, everyone cared, everyone felt the family's pain – and everyone wanted updates.

It could easily have become unmanageable, fielding multiple phone calls a day, answering the door repeatedly, but they hit on an answer: create a Facebook support page as a vehicle for news, and a place for people to show support. Susan couldn't get that off the ground without a profile of her own, so her friend Emily started Friends of Jonnie Sue, a closed Facebook group. Its membership quickly swelled into the hundreds, each new member personally vetted by Emily and Susan to ensure that they had a real connection to the family. Eventually, Susan got her own Facebook account and became an administrator of the group alongside Emily. In many ways, this hub to deliver bulletins about Jonnie Sue, and to receive expressions of caring from the 700 or 800 people on the group, was a godsend. 'The support we received was, for the most part, so uplifting. It gave us a lot of comfort to know so many people were supporting us, thinking of her, thinking of the two grandsons. So it was a great way to communicate things,' Susan said. But it wasn't always easy.

'There was almost a low-level expectation that put stress and anxiety on myself. You know. "I need to update the support page",'

Susan explained. 'In an odd way, I felt that responsibility to do that, because people were so kind, and they'd been so giving . . . . But I had to be very careful about the things that I posted. Everybody's listening to what I'm posting on there. The grandsons were also very active on Facebook, and I had to . . . make sure that anything that I wrote had already been communicated to them. It took thought. I couldn't just get on there and say, "Today she had a drainage tube put in, and she's very ill from the treatment, throwing up all night." You just can't say the real things that sometimes you want to say, because it's very sensitive to them. And I felt a sense that I had to try to be uplifting too . . . as time went on, it was harder and harder for me to present that positive outlook to the public.'

Susan also felt that she needed to check the posts with her daughter, the subject of all these reports and the nexus of everyone's concern, but sometimes it was too much for Jonnie Sue to manage. 'It wasn't that she wasn't interested, because I think she was. But she didn't want to be the one having to be responsible, if that makes sense.' That responsibility fell to Susan and Emily and, working together, it was manageable. But then John became ill. For a while, concerned members of the community received updates on both father and daughter via Friends of Jonnie Sue; after a while, updates on John shifted to Susan's own profile page. When Emily began dealing with her own cancer treatment, Susan said, 'That was just a kind of unique triangle right there that happened. I say that my daughter and my husband were beloved by the community, and she was as well. She was a unique teacher. I just can't stress that enough. I don't even have the words to describe the outpouring.'

It was an outpouring that, predictably, found a home on Emily's own support page, of which Susan also became an administrator. There's another word that Susan used a lot in her interview: 'flood'. Floods of support. Floods of condolence. Floods of grief. Floods that washed over the personal Facebook pages for Jonnie Sue, Susan and Emily, over the support pages for the people who were ill and who died, all of which eventually fell to Susan to manage. The thought of taking these pages down – or even memorialising Jonnie Sue's personal Facebook account – doesn't ever seem to have been a serious consideration. First of all, they're places where Susan herself can go. 'Instead of visiting a gravesite, people visit a web page,' she said, 'and say the things you would say back in the old days over a gravesite.' Furthermore, they're clearly important to so many other people, and Susan derives solace from seeing how other people continue to visit and remember – except when it gets to be too much.

The particularly painful season starts with November, the month of Jonnie Sue's birthday, the month when her unmemorialised profile triggers everyone on her list to send their birthday wishes to their lost friend, daughter, mother, sister. People post photos of her, some of which Susan has never seen, sometimes tagging her as well as Jonnie Sue so that the images pop up when Susan's not expecting them, taking her breath away for a moment. Later that month comes the death anniversary for John, cremated on Thanksgiving Day, and on Susan's own page the memories roll in. Next up are the death anniversaries of both Emily and Susan's son-in-law, both in December, the remembrances juxtaposing oddly with all the online Christmas cheer. 'You know. The holidays. People post photos of their happy families,' said Susan. 'Their families that are whole and not broken. It's all a huge string of triggers.' It's February before she can come up for air, for Jonnie Sue's death anniversary falls in January.

Engagement in social media can be hard to calibrate – if you're off, you're completely out, and if you're on, you get it all. If you adjust your settings, preventing people from tagging you, unfollowing the people you don't want to get updates from, disabling the 'on this day' photo function, you might be able to avoid triggers, but you expend energy, and you might lose out on the things you do want. There's more of an incentive to stay engaged if you feel that others are counting on you. This particular year, though, Susan didn't have the energy for it all. As the nights drew in and the leaves fell, signalling the start of the hardest season, she made a decision.

'It was late October,' she said. 'I thought, I can't do this another year. I can't bear to do it. I do OK as far as my grief, and being able to cope, but it's just excruciating sometimes. So I disabled my personal

Facebook account from the end of October. I'm gonna reactivate it, but I'm just going to wait until February. Maybe. And get back on. And that's the first time that I've done that, since the deaths.' She needed a break from her stewardship, and eventually, she knows that it will need to pass to someone else. Although there's no guarantee they will take the same approach, she presumes the torch will pass to her grandsons. Preserved on multiple devices, one of which is a clearly marked USB drive, she's stored all the information they'll need: a file entitled 'Passwords for Mom's stuff'.

In the October before I spoke to her, for perhaps the first time, Susan had taken a step to prioritise her own grief and her own needs over the responsibility she felt so keenly. It wasn't necessarily a responsibility that she felt directly to Jonnie Sue, or John, or Emily. It was to all the other people who had loved them. While Susan steps back from looking after those digital legacies, though, she's letting another part of a digital legacy look after her. Every night, for over forty years, Susan lay her head on the chest of the man who was the love of her life, her childhood sweetheart, and let the sound of his heartbeat carry her off to sleep. As he lay dying in a hospice, Susan knew he didn't have long. Feeling a rising sense of panic, she asked a nurse if she could help her capture something that felt particularly important to her. And so it is that now, each night, using her MP3 player or her smartphone, Susan falls asleep to the digitally recorded and preserved sound of her husband's beating heart. Her family may have always been at the heart of the community, a community to which Susan feels both a gratitude and a responsibility, but for now, online at least, and at least until February rolls around again, the grieving community will have to look after itself.

Ω

Sometimes what needs care and attention is not the digital biography, the consciously uploaded, visible material, but a dead person's digital archives: their emails, message histories, documents, photographs, and assorted files that were never intended to be widely shared. When

Vered's older brother Tal died in 2011, it was sudden. There was no long illness, no rallying around of the community, no culture of expectation developing over months and years that his younger sister would update people on what was happening with him. Instead, he left his apartment in Israel one morning, expecting to return home at the end of the day, and was killed in an instant, struck by a car.

Vered was faced with a different set of circumstances from Susan, but in managing her brother's digital legacy, she still felt a certain amount of responsibility to her brother's circle of friends, family and associates. 'He was an amazing photographer,' she said. 'And several people approached me and said he took pictures of them, or of their motorcycles, and he didn't get a chance to give them these photos.' Vered wanted to help, but for a start, going into Tal's apartment felt incredibly difficult. Powering up the computer that had last been used by her brother was hard too. Locating the photos in question from amongst the thousands of files on his laptops, based on sparse information such as names, approximate dates, and photo content, was another level of challenge, made easier by how well organised Tal was. 'I managed to find everything that they asked me for, and they were so thrilled. I felt like I was giving such a huge gift, you know, by locating those pictures . . . the last picture that he took of them,' Vered said. 'Giving to them something that was so important to them, I didn't mind spending time looking.'

That wasn't the end of it, however. Other people that were central in her brother's life needed things as well. Tal's best friend asked if Vered could delete her email correspondence with him, and Vered was happy to oblige, partly because she believes in the sanctity of private correspondence. 'I mean, what I say to my best friend, and what my best friend says to me, is only between us. No one else should be reading it,' she explained. Unlike Tal's photographs, his emails were housed on external servers and locked away behind passwords, so she wasn't sure how to go about this mission. Again, however, she was determined to try. 'I use this phrase, "fold". I needed to fold his life neatly behind him. And what we did with his apartment, his physical stuff, I felt like I should be doing with his non-physical stuff,' she said. It was easier said than done.

The process of hacking into her brother's emails involved a complex juggling act of impersonation and detective work. Swimming through unfamiliar waters, wary of any stroke that might sink her chances of accessing her brother's data, she proceeded with caution and surreptitiousness. 'I didn't know what I was doing. I didn't know if they had a policy – I didn't check. I felt intuitively that I had to write to them as if I *was* my brother, saying, "that's my account, I need access to it." Intuitively I assumed that if I would say, I am his sister, and he is dead, my chances of regaining access to it would be a lot leaner. I didn't understand that they would have been nonexistent.' She started with his main email account, but the provider asked her security questions to which she didn't have answers. Hoping to strike lucky, she asked his friends if they could guess the correct responses, and this strategy paid off – she was in. Having got into that email account, she went to the other sites to which she needed access, hitting 'forgot password' and resetting those passwords through the email to which she'd already gained access. Conscious that this work-around violated the one-account one-user terms and conditions, she didn't want to tell me what kind of email account Tal had. Her email was with the same provider, and she was afraid that disclosing the details to me and having them published might result in her losing her own accounts. It took months, but ultimately she was able to do what her brother's best friend had asked. One more item had been neatly put away. But there was more to do.

During the period of time that Vered was going through all of this digital material, Tal's adult children were agitating to get hold of their father's laptops. 'They just wanted his computers to themselves as soon as possible,' said Vered. 'They were in their early twenties at the time, their father was just killed . . . . they couldn't understand why it was taking me so long.' Part of the delay was connected with how psychologically and emotionally impactful Vered was finding her task. She felt the pain of her brother's children, and she smarted from their anger, but she couldn't give them the laptops yet. She had a job to do that nobody seemed to comprehend, and she persisted with it, however isolating and draining she found it. She asked her best friend to stay in the flat with her while she worked, close by but not in the same room, affording her privacy but still providing a sense of support, and she put Tal's children off a little while longer. This task would simply need to take however much time it took. As impatient as they were, Vered felt that she was actually discharging an important responsibility towards her brother's children, and perhaps towards their father as well. She didn't feel that the laptops and the accounts were ready to be handed over to them.

'My brother was fifty-five and a half years old,' said Vered. 'He was divorced. He had his own life. I didn't feel it was right to just give his computers as is to his children. He was single.' Her tone took on a certain significance, and she ducked her chin, raised her eyebrows, looked at me to see if I grasped her meaning. 'What I did was, I arranged his emails by the name of the sender, and if it was a female name, I opened it long enough just to glimpse and see if it was professional or personal. If it was personal, I deleted it without reading it. If it was professional, I left it as is. Because, you know, I felt like when you go out on a date, and you introduce yourself to someone, no one should be reading this but you and this person.'

I think of Susan, saving for her grandsons, not just Jonnie Sue's Facebook account and support pages, but the content of her email and Facebook Messenger as well. Just as Vered didn't read her brother's emails past the first line or so, Susan said she hadn't read Jonnie Sue's private correspondence, but she didn't feel any compunction in handing it over in its entirety to her grandsons some day. 'I don't think there would ever be anything in there that would be inappropriate for them to see,' Susan said, with confidence. If the boys eventually wanted to read them, that was their choice. Vered was taking no such chances. She was convinced that she was doing what needed to be done to safeguard not just Tal's privacy, but his children's experience as well.

Years ago, at a conference about death and dying, I was struck by a presentation about something I'd never thought about before: how the personal property of disaster victims was returned to their family in the UK,[232] and how these efforts were supported by a whole industry that I never knew existed. After a plane crash, a tsunami, an explosion, or a terrorist attack, the victims' personal possessions are photographed, and the images arranged into catalogues for identification by relatives. Armies of laundry workers clean, mend, iron and neatly arrange the clothing of the dead, packaging it up with other personal effects, and safely return it to the bereaved. Grieving families receive a tidy parcel, not a tangled mess. Vered, sitting with her brother's laptops in his quiet apartment, felt she was doing the same with his digital material: carefully, painstakingly folding her brother's life neatly behind him, and only then releasing it into the world.

Listening to Vered's story, I think about all the stakeholders there are in grief, and all those situations in which stewardship of the dead is contested. On one hand, I can understand Vered's perspective and can put myself in her shoes. At the same time, I think about Tal's kids, to whom I've never spoken. Would their aunt's ideas about what was appropriate have differed significantly from theirs? Some might call Vered's work 'scrubbing' – an infinitely more complex, digital version of my mother's attempts to render the swan memorial lily-white again. Ultimately, knowing Vered, I'd be inclined to think that if her actions were a kind of scrubbing, it was relatively benign and certainly well intentioned. There have been far more nefarious examples of this practice, carried out by people who had no doubt in their minds that the right to craft and perpetuate their loved one's digital legacy – or erase it altogether – was theirs, and theirs alone. Whether or not they were legally and morally correct in this assumption depends on your perspective.

Ω

Tom Bridegroom and Shane Bitney Crone had been together for six years. They co-owned a house, ran a business together, adopted a dog. By all accounts they were soul mates, completely in love. Under different circumstances they would have married, but in 2011, it was against the law in California, where the couple lived. Tragically, as Shane puts it in the film that he made about their life, Tom would never get to fulfil the promise of his last name. When he was killed in a freak accident, falling from the roof of a building while doing a photo shoot with a friend, his partner of six years was denied admittance to the hospital. He wasn't classified as 'family'.

This was the first of many painful marginalisations, not the least of which was this: Tom's parents requested the deletion of their son's Facebook profile, which encompassed the record of his life with and love for Shane. If erasing this element of their son's life were the intent behind the action, it is shocking and unsurprising at the same time. Tom's mother and father were from a conservative area in the Midwestern United States, and their church was unequivocal in teaching that homosexuality was a sin, an offence against God. Both online and off-, religious communities of grievers establish and mutually reinforce – in both overt and subtle ways – what they see as acceptable narratives about the deceased's life and death. To deviate from that narrative can mean isolation at the time you most need support. Excising online material that was dissonant with their beliefs may have been the only way that Tom's parents could manage, in their grief. With material like the Facebook profile gone, they could go on to build an offline, locally acceptable, alternative durable biography for Tom, one that fitted comfortably within their church community.

Groups of grievers reinforce and police 'acceptable' views about the deceased's life and afterlife in online environments, too. Layla, who grew up in a fundamentalist Christian sect, told me about a Facebook group of which she is a member, dedicated to commemorating those members of the church who pass away.[233] In this particular church, Layla explained, there's no notion of Heaven, angels or souls. People die, their bodies go into the ground, and they have no consciousness or awareness until the great mass resurrection, when everyone who has ever died will have the opportunity to know God, and to see one another again in the Kingdom if they accept Him. These views were constantly reinforced in people's comments on the Facebook page: there were no mentions of any Heaven, but plenty of remarks like, 'It'll be good to see them in the Kingdom' and 'We look forward to a glorious day ahead.' Some former members of the main branch of the church, however, had decided that they *did* rather like the idea of Heaven. That led to them splintering into alternative sects of the church at best or being excommunicated at worst. On this particular memorial page, though, the party line was held. 'From what I can tell in the memorial forum, they show a certain amount of respect for those who still do believe,' said Layla. 'If someone says, I look forward to seeing you in the Kingdom, or whatever the phrase might be, no one is going to jump on board and say, yeah, *that's* never going to happen, you're full of it, or, this person's already in Heaven ... there's none of that. There is a whole separate Facebook page for people who are non-believers to go and gripe about their [church] experiences! They tend to be nice on the memorial page.'

In his bereaved partner's view, however, there was neither nicety nor negotiation in Tom's parents' decision. On Tumblr, Shane posted the full text of his speech at Tom's memorial, a commemoration that was held partly to make up for his being banned by the family from Tom's actual funeral. 'Tom was an avid supporter of social media,' Shane wrote, 'and believed that information should be truthful and easily accessible, which is why it's so painful that his immediate family had his Facebook page deleted. He updated his Twitter and Facebook accounts regularly and shared his life openly and honestly with those he loved.'[234] Amongst those he loved most, of course, was Shane, a fact that his family seems to have been at pains to efface from Tom's legacy. As one contributor to Shane's film *Bridegroom* put it, 'they erased it from the history books'.[235]

As members of a religion with particular views on homosexuality, Tom Bridegroom's parents might have thought that curating their son's lasting image in this way, excising any material that they believed represented weakness or sin, was a responsible and eminently defensible act of stewardship, the only thing that was imaginable in their world. In 2011, as next of kin, and as per Facebook's policies, they had the right to make his social media profile, his record of his real life and his real love, disappear. And they did. 'That's the stuff that keeps me up at night,' says Jed Brubaker.

Jed Brubaker works for Facebook.

Ω

I was mildly surprised to be speaking to Jed, having been pessimistic that I'd ever find myself face to face with anyone possessing direct Facebook connections. My earlier conversations with post-mortem privacy specialist Edina Harbinja had lowered my expectations. 'Facebook, sadly, is quite closed,' Edina said, shaking her head with an air that suggested I shouldn't waste my time. 'They don't really want to collaborate with researchers.' I had her words in mind when I first wrote to their press office, so I called myself a 'journalist' rather than an academic or researcher. Despite this canny strategy, I never heard back from them. Maybe this is a bit what it's like for the families and friends of deceased users, I thought. I'd spoken with so many of them: bereaved people with axes to grind, people who'd sought voice-to-voice contact with someone at Facebook and who said they didn't have a snowball's chance in hell. Or maybe it wasn't usually this difficult, but I was simply getting lost in the shuffle of recent events. The Cambridge Analytica data-sharing scandal was fresh, and Mark Zuckerberg had just been testifying before Congress. That was big news, and hundreds of journalists and writers must have been getting in touch with the company every week.

But then I thought of Jed. He and I were both academics, studying the same things; we were aware of one another's work and knew people in common. Even if he did get back to me, though, I assumed he'd either give me superficial, canned responses, or he'd have a non-disclosure agreement zipping his lips. To my surprise, Jed wasn't just able to share his insider knowledge, but he was entirely willing and disarmingly transparent. His relative freedom to spill the beans with me was down to the fact that his work with Facebook fell under the auspices of the Compassion Research Team, an initiative started at the company in 2011 with the mission to 'ease difficult moments and enhance well-being in people's lives'.[236] Compassion's philosophy was that there wasn't any value or point in rendering good work invisible.

'In general, the rule I get from Facebook is don't talk about any-thing,' Jed admitted. 'But the rule with the Compassion projects has been, talk about whatever you want. Other Internet companies may take inspiration from this, and if they do, that's not an intellectual property issue. That's just going to make the Internet a better place.' And whereas Edina had been cynical about Facebook's willingness to collaborate with academics, Jed's perspective was different. When he talked about how Facebook actively reaches out to academics, seeking collaborations with them, he should know: Dr Jed Brubaker *is* one of those academics. At least, he is now. When Facebook first reached out to him, he wasn't a fully paid-up member of the academy just yet. He was still studying towards his PhD in Information and Computer Sciences and writing up his doctoral dissertation, *Death, Identity, & the Social Network*,[237] and he had been in the running for a 2014 Facebook Fellowship, a grant initiative for talented doctoral candidates carrying out relevant research in computer science and engineering.[238] While Jed didn't make the final cut for a fellowship, his candidacy paid off in a different way: Facebook became more aware of his research on death and social networks, and they realised that they needed him. The Compassion team wasn't happy with how Facebook was managing dead users' profiles. They knew that bereaved people, users and non-users alike, were experiencing far too many difficult moments, that there were various ways and places in which people were encountering 'pain points', as Jed referred to them at one stage.

Jed knew that his knowledge could make a difference and, although he was concerned about becoming seduced into an industry job before he'd finished his doctorate, he reckoned that there wasn't too much danger of that – a couple of weeks of 'knowledge transfer and whatnot' would surely do the trick. His interactions with Facebook en route to his eventual doctorate ended up being rather more intense, involving week-long trips to the Menlo Park Facebook HQ every few weeks, and an endless stream of meetings with engineers, designers, policy people, lawyers and members of the Compassion team. He and Vanessa Callison-Burch, who was then a product manager at Facebook and who remains one of Jed's personal heroes, were the driving forces behind Facebook's current stewardship model, including the launch of Legacy Contact in 2015. Although he's now an assistant professor at University of Colorado Boulder, and Vanessa is studying to become a

Buddhist chaplain, they're both still collaborating with Facebook.[239] He told me that, in the coming months, the summer of 2018, the two of them would be going on retreat to think about the next steps. The inspiration and challenge driving them, he said, would be the same as it ever was: 'How do you care for people . . . especially with the hall-of-mirrors effect that grieving can cause?'

How do you care for people? The very existence of that question as a central concern represented a significant change in position, Jed said. In the beginning, Facebook didn't exactly regard itself as a steward of dead people's memories or bereaved people's experiences, and their policy when one of their users died was purely dictated by their legal obligations. Issue: a living person we once made a user agreement with is now a deceased entity. Solution: delete. Jed shuddered slightly, describing the history of deletion as 'horrible'. It was a short chapter in the network's history, though, for the catalyst for change occurred in Facebook's early years: the massacre at Virginia Tech in April of 2007, in which over 30 people died.[240] 'People reached out to Facebook and said, these spaces are memorials, please don't delete them,' explained Jed. The policy of automatic deletion was duly dropped, a move that had both ethical and pragmatic implications.

Ethically, the change in policy meant that for the first time Facebook was accepting moral responsibility for the preservation and maintenance of memorials on the site – not just the memorial groups that were proliferating there, but the digital identities persisting in the form of individual profiles. Pragmatically, it meant that they were accepting practical responsibility for all that as well but, for all the reasons described in previous chapters, they had their work cut out for them. 'There were so many moving pieces . . . it was such a nascent thing,' said Jed. In the year that Virginia Tech happened, the site had 58 million monthly users.[241] By the time Jed was called on to help Facebook think through design and policy for memorialisation, the need to develop better processes for deceased people's data had intensified. At that point, over 1.23 billion people – twenty times the number of users in 2007 – were logging on regularly.[242]

Even if the number of moving pieces hadn't increased in seven years, at the very least, the death of users on the site was affecting far more people, for good or for ill. When he arrived at Facebook's headquarters, some of the first meetings Jed had were with the members of the Compassion team. He described these as being inspiring dialogues with people who each had a deeply felt reason for being there, and who were passionately dedicated to the project as an almost-sacred duty. He also had a lot of meetings with lawyers, which he described rather differently. 'When I would have arguments with the lawyers ... "arguments", well, I'm being a little bit tongue in cheek. Really they felt like they were stripped right out of a Hollywood movie. They were vigorous intellectual debates while walking laps on Main Street,' he said, referring to the central thoroughfare of Menlo Park. 'My goal in those conversations was, we've done the research, and we're hearing that what's there kind of sucks. It needs to be improved. We've taken a human-centred approach to doing design research, and here are what would be better for them. How can we make that work in law?' On the chopping block for dissection were all of the legal challenges detailed in the earlier chapters of this book, questions that drove up the step-counter as Jed and the lawyers paced up and down Main Street. Who owns this stuff? Who has the rights? Who is allowed to do what? Do the laws of traditional inheritance stand up when you're talking about digital artefacts?

If you've been keeping up thus far, you'll know that there remain few clear answers to any of those questions, and meanwhile the need is outpacing the far statelier progress of legal reform. So Jed knows this much. If he, Vanessa and the Compassion team were going to wait around for absolute legal clarity before they moved forward, they'd still be in the same place they were in 2007. Maybe by doing the necessary research, and by building and implementing policies that they felt worked better for a digital age, they could help *shape* the laws of the future. Maybe they could actually lead the way, not just in terms of developing template contracts and procedures for other Internet companies to follow, but also in terms of changing how we legislate, regulate and *think*.

'We as a global culture need a shift,' said Jed, and if anyone can force the hand of global culture, maybe it's the world's most influential social network, its power summed up in something Jed expressed more than once: 'Because we're Facebook, we can.' So, with the best will in the world, they're not waiting for permission. 'I am not daunted by the legal issues,' said Jed. 'The law is here to serve us, not vice versa. Clearly, it's a policy nightmare . . . . we acknowledged it was a mess, but we were like, *let's take care of what we can*. We've been given this really privileged opportunity to do something that can not only help people on the platform, but . . . given all the problems of inheritance, how do we actually put something out in the world that shows an alternative?'

The alternative he's talking about, in part, is the idea that next of kin shouldn't have automatic decision-making power when it comes to deceased people's digital footprints. Legacy Contact embodies that alternative, giving both the deceased and their designated trusted contacts the chance to express, and possibly even to enforce, the deceased's wishes for what happens to their Facebook profiles after they die. Jed admits being deeply affected by stories of the pain caused when families of origin have scrubbed profiles, like the tale of Tom Bridegroom and Shane Bitney Crone. 'Do you privilege family of birth, or family of choice?' Jed asked, and his take on that is pretty clear. Next of kin, he argued, have other memorialising spaces. They might think of the Facebook profile as just one more thing to shut down alongside the bank accounts and store accounts. For people with looser ties, though, a social networking site might be the only place where they can connect with the deceased. 'The people who are most able to make choices are the least invested,' Jed said, 'and the people who are the most invested are least able. There's these asymmetries that happen that I think are really important to pay attention to in a social media space like Facebook; asymmetries that are not well represented or even well accounted for.' Those are the asymmetries that Legacy Contact was meant to rectify.

Shane Bitney Crone, for all practical purposes Tom's widower, is acutely aware of that asymmetry. Having been banned from his partner's funeral, Shane eventually flew to Indiana to visit Tom's grave. When he located the marker, he discovered that the Bridegrooms had bought their own plots and erected their grave markers on either side of their son, with barely a finger's breadth between the headstones.[243] Shane had gained physical access to the cemetery, but he felt symbolically excluded. He flew back home to California, returning to a state where the Revised Uniform Fiduciary Access to Digital Assets Act was eventually enacted in 2017.[244] Had it existed in 2011, and had Tom designated Shane as his Legacy Contact on Facebook, Shane would have had legal authority over Tom's profile. Acting as a kind of joint fiduciary with Facebook, the host of Tom's memorialised account, he would have had the legal power to prevent the Bridegrooms from deleting the account.

Depending on your perspective, that would have been either a victory or a defeat. I suspect that I know what the Bridegrooms, or Sharon, or Rachel might have felt about laws barring them from access to or control over their children's digital material. But when Jed said that the law is here to serve *us*, he wasn't talking about next of kin. He was primarily talking about the deceased onetime account holder, and the people who were important in their lives, those to whom they would have most likely entrusted their lasting legacy. And if that deceased hasn't set their Legacy Contact preferences – well, given all the asymmetries and the potential for pain, he reckons that preservation is a safer bet than deletion.

Even so, it's not that Jed has no sympathy for next of kin. We discuss the Hollie Gazzard case. 'I do recall that scenario,' Jed said. 'I feel for those families . . . . I've been worrying about things like that. I've been thinking a lot about what happens on Timelines when content is preserved that could be very painful, the cyberbullying, gang stuff that resulted in a homicide . . . . the temporalities in [Hollie's] scenario, and the nuance right around the point of death.' Jed clearly knew how difficult it would be to argue that retaining her murderer's images on her profile would have been Hollie's wish, and he was keen to emphasise that he knew that work was far from finished. 'Every death is exceptional,' he said. '*We're not done.* There is stuff in the works that will make people's experiences better.'

In service of that, Jed and his colleagues keep sifting through and reshuffling the long list of options and evolving priorities. 'Where do we focus? Is it in something visible? Is it in the policy behind the scenes? Is it more about what memorials look like? Is it more about what Legacy Contacts can do? Sometimes you can go about fixing what's in the box, but you can also change the box. Or add a box.' Maybe not too many boxes, though. Jed and the Compassion team are on a quest for the perfect design and policy solution, a solution that will not just increase the numbers of users who express their wishes, but that will help Facebook care better for the bereaved. He knows that those are people in a state of shock and grief, likely to be easily confused and flummoxed by granularity, in need of care and guidance. 'The metaphor of the funeral director is not far off, right?' he said. 'We don't ask the bereaved to go and embalm their loved ones. A lot of that complexity [can be] taken away. Call it an example of good design, so that people make choices that meet them where they're at.'

While Facebook didn't include anything like 'funeral director' in its original business model, and while caring for the final disposition of billions of digital citizens may be a challenge, Jed clearly has faith that good design can make it possible. In any case, he isn't prepared to duck what he sees as a sacred ethical responsibility to Facebook's population of over 2.38 billion monthly users[245]. In April 2019, the small team behind Facebook's memorialisation policies assumed that mantle of responsibility once again, announcing that they were 'Making it Easier to Honor a Loved One on Facebook After They Pass Away'[246]. One major structural innovation was the introduction of a separate 'Tributes' page linked to each 'Remembering' account.

The new raft of memorialisation features was designed to make the site 'even more supportive', the press release said, but I had reservations about some of them. Using AI to detect profiles of deceased users and to stop birthday reminders from those accounts, for example, would only be helpful to those mourners who *wanted* birthday reminders to cease. I remembered Sharon, for example, who was devastated when birthday reminders for her daughter Amy ceased. Furthermore, the granting of greater editorial powers to Legacy Contacts would only be favourable if a Legacy Contact were prepared and equipped to moderate a Facebook profile for years to come.

So here we are, in a position where big tech is telling us what the easiest and best ways are for us to honour our loved ones. One might argue that this is a moral and psychological judgement that – particularly given the idiosyncrasy of grief – they are not best placed to make. Nevertheless, Jed hopes that Facebook will be an industry leader for other Internet companies, companies that have no less obligation to their users, but that may need different solutions. 'What I would really love in an ideal world is for more companies to be having these conversations. I want to *know* what the right thing is! I want to know what the appropriate approach would be for Amazon, or something really mundane, like Google Drive,' he said. 'I want to know!'

But what happens, I wonder, when the commercial interests of any company conflict with their moral and ethical ones? For Facebook itself, wouldn't a tipping point be reached when the dead would exceed the living on the site? Making the baldly cynical assumption that any corporation is driven primarily by profit, I wondered: if Facebook *does* still exist at that point, what financial incentive would there be to maintain the business any longer? And in the meantime, is there anything to gain financially by maintaining the posthumously persistent digital identities of millions of dead erstwhile users on their servers? There's no advertising on memorialised, deprovisioned accounts. Is that really just chalked up as a loss on the balance sheet?

I had turned it over in my mind, and I thought I had a plausible hypothesis. If you're a long-time active Facebook user, a significant amount of your life will be archived there. If you use the site as a major conduit for connections with friends, your relationships and conversations will be archived there too. And the longer you stick around as a user, the more people you know on the site will die. The more people become aware of the value that online memorials hold for many grievers, the more likely it is that profiles will be retained, rather than deleted. Think about what a powerful incentive that is for living users to stay on Facebook – if you delete your account, you lock yourself out of the cemetery. You take your own shoebox full of letters and photographs, and you set it alight.

Jed Brubaker's passion is real, and the ethical sensibilities he expresses feel entirely genuine. He's full of admiration for the caring and commitment of the Compassion team at Facebook. I had some reservations, therefore, in putting my hypothesis to him, but even so, I felt that I had to ask him. Taking a deep breath and assuming the role of devil's advocate, I asked whether all this talk of a moral obligation might just be a fancy window dressing behind which baser motives lurked. Was the memorialisation of profiles merely one of Facebook's many methods of retaining living users, those who cherished and mourned the dead and valued their digital artefacts? As the years pass, and as the online cemetery fills with loved ones, might this not be an increasingly commercially meaningful factor in maintaining their business?

Jed paused before he answered, and I wondered whether I'd caused offence. 'To date, there's pretty good evidence that there are not [commercial incentives],' he said, emphatically. 'If there are . . . they're pretty crappy ones. There's not a lot of money to be made in cemeteries. There's way more money to be made in amusement parks.' And what about when the digital worms turn? I asked. When the cemetery grows too large, might Facebook close up shop and move to greener pastures, focusing on building new, more lucrative amusement parks? 'I'm sure that there is someplace where the rubber meets the road,' replied Jed. 'I'm sure there's some reality check. I'm sure that this might change over time. But money's not a factor here.'

Jed is right about many things, and one of them is this: there's not a lot of money to be made in cemeteries. That's more the case today than ever, and the digital age may be one of the reasons for that.

Ω

When it comes to looking after and memorialising the dead for evermore, it's no longer a foregone conclusion that cemeteries will be a major player, a central focus for memorialisation. Business is tough; fewer people are seeking immortality in stone. In an industry suffused with tradition and history, Jon Reece and his colleagues have tried to be as forward-thinking as they can when considering how they can inspire people to keep on using physical cemeteries. As the president and general manager of Seattle's Quiring Monuments, Inc.,[247] Jon has a vested interest in fighting what seems to be an increasingly uphill battle. He listened with interest to my story about the swan and gave me a potted history of the different headstone materials Americans have used over the years in their efforts to achieve maximally lasting memorials. Early white settlers on the East Coast primarily used thin slabs of engraved slate. 'At the time they thought, this is a permanent record, but you can't see anything. There is no record of who is there, it's gone,' Jon said. Then people adopted marble as their material of choice, but particularly in urban areas where there's a lot of pollution, marble is melting; the lettering and sculpting on memorials like my grandparents' swan is gradually dissolving. In more recent times, cheap and easy concrete has been used as a base for granite or bronze markers, rendering monuments vulnerable to damage and unsafe to visitors as the concrete crumbles away. 'So we have made mistakes, and we have tried and failed in our intent to preserve the history of our loved ones,' said Jon. These days, it's pretty much all about granite.

Most recently, though, the dilemma seems to have changed. It's not about *which* stone to use; it's about whether there will be any stone at all, in any cemetery. Burial is expensive, and cemeteries tend to be highly regulated and nannyish about how monuments can look and what the inscriptions can say. 'People don't want to be told no, they don't want to have handcuffs, they don't want to have rules, they want to be able to express themselves,' said Jon. 'Cemeteries are saying, no, no, we don't like that. Don't do that. Control and regulation, in my opinion, are killing the cemetery industry.'

So the rise in cremation's popularity isn't just because it's cheaper, Jon thinks – it's about control and choice. The portability and divisibility of ashes means that people can keep the physical remains of their loved ones nearby, in memorials of their own preference or construction. They can decide their own rituals, make their own dispositions. On the West Coast around Seattle, he said, the cremation rate is about 75 per cent. He described how people scatter ashes to the four winds off Mount Rainier, perhaps the busiest cemetery in the state of Washington, or cast them off the decks of Seattle's ferries, supported by the ferry companies. If it's the ashes of a member of the armed forces or a police officer that are being sent into the sea, the ferry will pause, and a bagpiper will play. If people want a memorial, they have the Internet for that, and Jon recognises digital artefacts or memorials as also being about individual expression, affordability and control. 'With people mourning on Facebook, people felt like, hey, I can do this memorial on Facebook, and I don't need to do a marker any more,' he said. 'This was something that over a period of time would erode our business, so we thought that, with the digital age, we've got to try different things.'

In a bid to future-proof their business, and with an if-you-can't-beat-'em-join-'em spirit, this traditional stonemason set out to produce and market some kind of hybrid digital/physical memorial. Bridging the gap between the needs and preferences of digital immigrants and digital natives, they thought, might be just the ticket. At first, they looked into devices like the telephone receivers you pick up to listen to recordings at museum exhibits, but they didn't like the idea of feeling 'tethered', as Jon put it. As smartphones developed further, they explored using near-field communication technology, where a connection would spring into life as a smartphone user drew near to a diode affixed on a granite headstone. And what if there were a sign next to a monument, displaying a URL address that a visitor could tap in to access a memorial website? Maybe that would be too fiddly. Then came the eureka moment.

'Suddenly, QR codes appeared!' Jon recalled. 'And we were like, it's so easy! It just fell into our lap! We got started. We came up with a name – we wanted it to be a little profound to get people's attention – and finally we settled on "Living Headstones". We wanted a little bit of shock value, but not to be offensive.' The plan was to produce a simple plastic-substrate QR code, with a nice copper finish, black painted and laser engraved with a code that would link to a bespoke memorial website. They weren't software guys, Jon said, so they partnered with another company who could produce and maintain the memorial sites. The QR code could be affixed to any granite monument, and another easily produced if it were to fall off or become damaged by the elements.

On the launch of the product, Quiring was overwhelmed by a tide of local and then national media attention – National Public Radio (NPR), *USA Today*, the lot. 'I kept a list for a while and it became exhausting,' said Jon. Startled by the amount of exposure, they scrambled to prepare their business to be the next big thing in memorialisation. Other companies were copying their idea and engaging in one-upmanship, offering flashy ceramic and steel QR codes, or codes engraved directly into the granite. And then: the sound of crickets chirping. In fact, when I first got on the phone with Jon, seven years after this avalanche of media attention, he sounded every bit as surprised to be speaking to me as I had felt to be speaking with Jed at Facebook. 'Can I ask you . . . how did you find us?' he said, with the excited and curious, but rather tentative, tone of someone who is startled to discover that anyone still cares.

Although they had recently been discussing a package with the Veterans Administration cemeteries, Quiring generally only sells a handful of QR codes per month. Part of the reason, Jon thinks, is that people don't trust that the QR codes, or Quiring, or the memorial webpages that the codes link to, will be around for ever. He doesn't consider himself in any position to reassure them. 'They're like, how long will this be around? What if you guys go out of business? . . . . Well, what can you tell them? If the cemeteries don't need us, and we go out of business, I'm sorry, but I'm not going to host those websites any more.'

I don't mistake his pragmatism for coldness – he's telling it like it is. 'In a couple of decades, could be generations, people are going to be going, what is that weird thing with all the weird squiggly marks?' said Jon. I think about the monument companies offering QR codes etched directly into the granite. What a curiosity that alien-looking symbol will be to the people of the not-too-distant future – if anyone is still visiting cemeteries, that is. QR codes, the sites to which they link, URL codes, Facebook, the Internet as we know it – in twenty years, all may have died their own deaths. Marc Saner, the current proprietor of the online World Wide Cemetery, with its infrequent digital interments but its 100-year fund set aside with the hope of protecting the ones that are already there, is as philosophical about this as anyone. Even though the maintenance money is there, he doesn't know whether anyone will be available to migrate the WWC's memorials to new platforms as the technology evolves. 'We have baby Internet,' he said. 'We are clueless. It's like the printing press in the 1500s, 1600s. We have no experience. We don't know yet what will be done. We have to culturally evolve in discussing what we want to do with all this.'

Ω

Gary Burks isn't particularly interested in culturally evolving. He guffawed when I called him a digital immigrant and suggested that 'digital ignoramus' might be more suitable terminology. He's more at home with the massive grave registers in the records room at the City of London Cemetery & Crematorium, bound in tooled red leather, full of copperplate writing, and fastened with straps and metal buckles. His daughter had to drag him kicking and screaming into even using WhatsApp. Although fascinated by the topic of my book, he's a bit scared by it too, reiterating multiple times how terrifically pleased he is to have very little digital footprint himself. Gary's made a significant stamp on his physical environment, however, having served a long tenure at this East London cemetery. He decided as a teenager that being cooped up in an office didn't suit him, so he started out as a gravedigger and gardener, tending the flowers around the graves, mowing the gently rolling hills and raking the tree-lined avenues, and he worked his way all the way up to superintendent and registrar, the big boss. The 200-acre space he oversees is a Grade I listed landscape, full of historic monuments dating to the mid-nineteenth century, but it's an active burial place too, providing for a thousand burials a year. There isn't a QR code in sight.

'I've had several people come and talk to me over the years,' Gary said. 'Requests from people trying to *sell* them to me . . . but I've not had a request for a QR code or anything on a memorial from family. They worry the life out of me. I know this sounds like control . . . but it's not, honestly. We have to manage every instruction that goes on a grave in this place.' His approach may be an example of the legacy policing that Jon Reece spoke about, but Gary privileges serenity and inoffensiveness over the right to individual expression. He told me, for example, about one family that intended to inscribe 'Murdered by the Police' on a headstone, and another family that planned to immortalise the deceased in stone as a 'hairy-arsed bastard'. On another occasion, he was alerted by his crematorium manager that a family had asked for a gangster-rap track referring to rape and murder as the chosen piece of music sending the deceased off to their final rest. In all three cases, Gary made a judgement call and intervened, averting the risk of causing offence to others. And if something happens to an existing monument in his cemetery, it's easily handled. 'We were flagged up to an inscription that had been badly mangled by someone, and written in permanent marker was a very, very nasty piece of text,' he said. 'Really simple. I just put a bag over it. We contacted the family about it, and we kept it covered up until the family were ready to do something about it. That's something I can do, because it's a physical thing.'

With online material, Gary reckons, control goes right out of the window, and he doesn't like it. Gary wouldn't know what kind of material a QR code linked to. What if it linked to something inappropriate? What if someone pasted another QR code over the top? As he emphasised so often in our discussion, as a 'digital dinosaur/ignoramus', he wouldn't even know how to scan the code to check. And what would happen when the technology became obsolete? 'So I've stepped away from it,' he said, unapologetically. 'I'm maintaining a cemetery in the classic sense of the word.' If someone wanted to know more about a person buried there, he reckoned, it was easy enough for them to use Google to find out more. Anyway, a hundred years from now, walking through the City of London Cemetery, the monument of someone buried in 2018 might not even be there any more.

Gary has the same concerns about the sustainability of cemeteries as Jon, but he has a different way of dealing with it. While Quiring is hoping to attract customers by using technology to 'bring your headstone to life, and share the memories of a loved one *forever*',[248] Gary keeps the cemetery's income stream steady by not trading in for ever at all. That's a departure from the Victorian sensibilities that held sway when the cemetery was founded. Victorians loved the notion of perpetuity, a word we don't use much any more. The permanence of final resting places was enshrined in England and Wales in the 1857 Burial Act, which said that to disturb human remains you'd need a damn good justification and formal permission from the Home Office or the Church.[249] In the City of London Cemetery's early days, all the grave plots, columbarium niches, and plantings in the memorial gardens were sold in perpetuity. 'I've got graves that we're maintaining "in perpetuity", planting plants, and we probably got a couple of pounds in 1902 to do it. It's not particularly good, in business terms,' Gary said.

Perpetuity was so bad for the balance books, in fact, that the London Cemetery Company went bust in the early 1970s, and the laws had to change. Now, just like in most European countries, but with a slightly longer possible lifespan, graves in Gary's cemetery are leased, not bought. The longest lease anyone can get is a century; as for the old Victorian graves, perpetuity has been drastically reduced to seventy-five years. When the lease is up, if there's no objection from anyone, memorials can be cleared. Headstones that are too damaged may be broken up into rubble, while monuments that are in better shape may be deemed Heritage Markers, spruced up and reinscribed for new tenants. Strolling the beautifully landscaped avenues of the City of London Cemetery, you won't see QR code panels, but you *will* spot multiple notices to the public, declaring the cemetery's intent to disturb human remains after a certain period of time.[250]

'Is perpetuity *right*?' said Gary. 'I'm not the one to say it isn't, but actually, perpetuity means that this cemetery will end up poorly maintained and not being able to offer services to local people.' What he's trying to convey is that it's not personal. It's not meant disrespectfully. It's just pragmatic: if you're laid to rest in London, or in any one of a number of mainland European countries, your physical memorial is likely to be cleared away after anywhere between twenty and a hundred years, depending on local laws and/or how long your estate wants to pay the rent. The promise of preserving

your legacy in perpetuity, however, is alive and well online. If Facebook will retain your memorialised social networking profile ad infinitum, if online legacy services will send messages to your loved ones for years after you're gone, and if other services will sort you out with a digital-twin avatar that will carry on socialising after you die, it won't matter too much if pieces of your headstone end up as part of someone's patio in twenty years' time, right?

But what if that screeching noise you're hearing is the not-so-distant sound of the rubber meeting the road, as Jed Brubaker put it? By 2025, someone told me, we're each going to be generating around 60 gigabytes of data per day, so much that it would only take a couple of days for the brand-new computer on which I typed this book to be choked and stalled by my personal data. With this looming capacity crisis, my source said, the threat of running out of space to store data will be an increasing challenge for every business, not just his. And his business? Ironically, it's Eternime, currently beta testing virtual immortality. Of course, immortality could be a pie-in-the-sky promise if the inexorable march of technological obsolescence, combined with mass culling of surplus data, produces a twenty-first century Dark Ages. In that scenario, millions of early digital-age citizens could be utterly lost to history.

But let's suspend our cynicism just a little bit longer. If Victorians dreamed of perpetuity, the New Elizabethans should be allowed to fantasise about it too. So, as Eternime's website says, who wants to live for ever? And how can we go about it?

Chapter Six
The Uncanny Valley

'What is science fiction?' I asked Google. I was preparing for my chat with Marius Ursache, founder and CEO of the Eternime app. Eternime had received a lot of press attention, but I didn't know that much about it, partly because it was still in the midst of a private beta. When I scrolled through the website, the functions that the app would perform looked and sounded entirely familiar to me, similar to dozens of other digital-legacy services that I'd seen. 'What if you could preserve your parents' memories forever?' the website asked. 'What if you could preserve your legacy for the future?' But then it veered into less well-known territory. 'What if you could live on forever as a digital avatar? And people in the future could actually interact with your memories, stories and ideas, almost as if they were talking to you? Eternime . . . creates an intelligent avatar that looks like you. This avatar will live forever . . . '

I felt my eyebrow going up and my brow furrowing, sensed the emergence of scepticism, a frisson of unease. Perhaps part of my discomfort was the awareness of my ignorance, for I freely admit that at that point I had no idea what an avatar was. I never fancied seeing that film where lots of the people were blue. I pulled up some of the news stories that, ages ago, I'd bookmarked about Marius's company. 'Eternime wants you to live forever as a digital ghost,' read one headline. The article went on to describe his inspiration.[251] Rather than things like spirit-guides, séances and Spiritualism, it read, Marius drew his ideas for virtual immortality from science fiction.

Not knowing my Asimov from my Heinlein, I suddenly recognised that I didn't exactly get what science fiction was either. It's possible that I would have been clear enough to know it when I saw it, but I wasn't sure that I had grasped the definition of it so thoroughly that I could explain it to anyone else. Luckily, if you're in the same club as me, you can quickly locate the online version of *The Encyclopedia of Science Fiction*.[252] Unluckily, it explains that the genre eludes easy definition and, according to the editors, we shouldn't hold our breaths waiting, as they don't believe we'll ever get a clear definition when the genre is so diverse. Emboldened by this shoulder-shrug, I gave myself permission to pick the definition that suited my purposes best. It's a synthesis of explanations offered by two acknowledged science-fiction greats: Philip K. Dick and Ray Bradbury.

Since 'science' is in the name, you might presume, as I did, that cutting-edge or futuristic technology would always have to be part of a science-fiction plot. Philip K. Dick didn't reckon that this was strictly necessary. What a science-fiction storyteller *does* need to do, he said, is set their tale against the backdrop of a society that we would recognise, for the most part; it should resonate with us, and feel plausibly rooted in the world we know, until the twist comes, and we see that something about the semi-familiar world portrayed is fundamentally different, novel, changed. Whatever that something is, it sets off a chain of events that couldn't possibly happen today, because our world doesn't contain that element. Like a cook who surprises and unsettles us by mixing a mystery ingredient into our favourite recipe, the science-fiction writer portrays a situation that we readily relate to, then chucks in something that gives us a 'shock of dysrecognition'.[253] Unlike Philip K. Dick, Ray Bradbury *did* insist that a science-fiction story feature a hefty helping of tech, but said it always includes a commentary on human society as well – a recognition of both our frailty and our potential. 'Above all,' said Bradbury, 'science fiction is the fiction of warm-blooded human men and women, sometimes elevated and sometimes crushed by their machines.'[254]

I now remember why I'd placed the interview with Marius into the 'get around to it later' category in my diary. As many psychologists do, I tend to privilege face-to-face interaction with warm-blooded humans and, anyway, science-fiction stuff has simply never been my thing. I didn't really understand the appeal of avatars and AI and androids, and I'd been having enough difficulty with getting my head around algorithms. But I couldn't put off our chat for ever. Perhaps it would help me relate to Marius and his app if I knew what kind of science fiction had inspired him, I thought.

When we did meet, the discussion was so fast-paced that I never had the chance to ask him which science fiction had got his creative juices flowing, but two things came up naturally in our conversation. I was unsurprised at his mentioning 'Be Right Back',[255] the 2013 episode that kicked off the second series of *Black Mirror*, a popular UK programme featuring lashings of 'what if?' futuristic dystopia. Not only had I seen the episode, but I must have had dozens of discussions about it over the years. 'Be Right Back' seemed to be a primary cultural reference point for anything to do with death and the digital, and people were always bringing it up when I told them what I focused on in my research. By contrast, Marius's other viewing tip, the 2004 film *The Final Cut*,[256] was completely unknown to me.

Ω

Alan Hakman, the solemn protagonist of *The Final Cut*, possesses the contained, professional, respectful mien of someone who works with death and final disposition. We quickly discover that he is indeed involved with the funeral industry, but his precise role within it is something new. One of his acquaintances, learning about his work for the first time, is fascinated and awestruck by what he does. 'You're like a mortician ... or priest ... or a taxidermist,' she says. 'All three.'[257] Alan's career and his craft are a synthesis of film director and funeral director; he creates feature-length biographical movies of people's lives to be shown at ReMemory services, the funerals of the future.

I'm reminded of *Beyond Goodbye*, a 2011 movie created by two filmmaker parents to memorialise their son, Josh,[258] and the memorial video one of my hometown acquaintances, filmmaker Greg King, compiled for his friend Jason;[259] both are available on Vimeo. Then there are the photo-montage memorials set to music and shown at funeral services, like the one made for another old friend, Stephen, still viewable years later on YouTube.[260] On one hand, Alan produces something akin to these phenomena, curating a selective, subjective summing-up of a particular person's life. On the other, because of the nature of his source material, his creations are an entirely different proposition. Early in the film, we begin to grasp what that source material is, through a video advertisement that's being shown on public transport.

'EYEtech introduces the ninth generation of the Zoe implant. From our family ... to yours,' intones the voiceover, as the advertisement displays a typical scene familiar to me from real-life 'digital legacy' websites – a smiling mother in a sunlit landscape, playfully swinging her child up towards a beautiful blue sky. Watching the ad, an expectant couple daydream about being able to afford this elite product. Only 1 in 20 of the population has one. But what exactly is it?

Eventually, it emerges that EYEtech manufactures a neurological implant, installed *in utero*, which stores everything that the audio and visual receptors of its host will ever perceive from birth until death. It's the ultimate life logger, infinitely more sophisticated, comprehensive and integrated than the real-life Narrative and Autographer wearable devices that were hitting the news around 2013. The implant is so undetectable that some people might not realise they have them, a fact responsible for one of the movie's primary plot twists. And why capture all of this information? Purely to ensure that the person can be commemorated and memorialised through a ReMemory after their death. The data has no practical utility while the person is alive, and the data from the implant can only be accessed when they die. 'You've made an important decision,' says the EYEtech representative on the welcome video for new customers. 'You've purchased a Zoe implant for your unborn [child]. What does that mean? Immortality. No longer do our most cherished moments together have to fade and disappear over time ... always and for ever, [your child's] experiences and adventures will be revisited, and relished, and that's what great memories are made of. Congratulations!'

Having chosen to call my child by the same name, I am immediately able to grasp why this fictional product is called what it is. Zoe: an ancient word, the name by which Eve was known in the Greek translation of the Bible. It means 'life'. But the amount of life footage captured by the Zoe implant presents professional 'cutters' like Alan with an extraordinary challenge. Assisted by sophisticated scanning and categorising software, but primarily guided by what the deceased's family both wish to remember and want to forget, Alan must select moments from across an entire life to produce the final cut: 100 minutes of carefully edited highlights, scored with an appropriately moving soundtrack, an emotive summary of an individual's legacy.

Numerous ethical issues are explicitly raised in the film, concerns that have been echoed off-screen in real life by critics suspicious of logging devices like the Narrative clip. An anti-Zoe activist challenges Alan on those sticky wickets: how might the Zoe implant change how we relate to others, threaten others' privacy, inculcate self-consciousness and doubt about every interaction? Alan has an additional ethical stumbling block, however, for the footage he views often contains criminal or otherwise reprehensible behaviour. With apparent stoicism, he does what he's always done, vaulting over this moral hurdle and scrubbing the nastiness out. He sees himself as a sin-eater, paid to remove the deceased's faults so that they can pass more easily into the great beyond, and so that mourners can be comforted with a soft-focus sentimental montage rather than cinéma-vérité. Not for nothing has he become known for his ability and willingness to produce lovely ReMemories for those who've led a less-than-exemplary existence. And that's what the grieving family members want, perhaps especially those who know their beloveds were less than perfect. 'I heard you were the best, that you would know how to handle Charles,' says one grieving wife, wearing widow's weeds for a rather horrible husband and making her expectations clear. 'I've seen ReMemories where the cutters were careless. *They have no respect for the dead.*' She knows that Alan, in possession of the entire record of her husband's life, has near-absolute control over how this man will be remembered. Using the 'guillotine', his editing machine, Alan can either elevate or crush the humans that have been captured on film.

The Final Cut readily stacks up against sci-fi criteria as set out by Philip K. Dick and Ray Bradbury. Alan's world looks very much like ours – there are the bookstores, the cinema, the advertisements on the bus. When Alan sits with a grieving family member, it sounds like he's doing what any eulogist, funeral celebrant or obituary writer would do, gathering information about moments and characteristics to emphasise in the sending-off speech. When mourners gather, dressed in black, the setting looks entirely familiar; there is the funeral home, the framed picture, the guest book. The graveyard is full of inscribed granite monuments. But then there's the shock of the new: a video panel flickers into life on the surface of the granite as someone draws close to it and, for all their invisibility, Zoe implants provoke societal changes that the manufacturer likely never anticipated.

The shock of dysrecognition was clearly too much for many reviewers in 2004. Although many of those reviews are no longer available in full (2004 was a lifetime ago in Internet terms), snippets from them are still on the review-compendium website Rotten Tomatoes, marked with a juicy red tomato for a positive assessment, or a sickly green splat for a thumbs down.[261] There is no shortage of splats. The *Atlanta Journal-Constitution* reckons that the conundrum posed by the film is worthy of Philip K. Dick himself, but the *Orlando Weekly* cannot suspend its disbelief in the face of such a 'preposterous' technology, employing an adjective that pops up often across the reviews. 'The premise sounds like the sort of screwy non-idea that a young film student might dream up while editing a documentary', says the *Orlando Sentinel*. 'Sheer nonsense', reckoned the *Miami Herald*, and the *New York Daily News* sneered at its 'weirdness'. 'Creepy', an adjective that Americans tend to use as shorthand for 'having anything to do with death', gets a good airing across the reviews as well.[262]

Audiences were no more impressed than reviewers, so during *The Final Cut*'s time on screens it barely scraped half a million dollars, ranking 249th on the list of US box-office takings for 2004,[263] a year in which the American public would have preferred to sit down to practically anything else at the cinema, to include foreign-language films. This was rather surprising, considering that the star of *The Final Cut* was Robin Williams, an actor whose involvement in any project usually spelled average takings of over $65 million.[264]

Ω

In nine years, a lot can change. In 2013, the 'Be Right Back' episode of *Black Mirror* essentially explored the same science-fiction question as *The Final Cut*: what if the data we generate in life could be converted into a kind of digital immortality when we die? Once again, we see a world that looks largely familiar to us. Cars look pretty much the same and travel on the roads, not through the air. When we first meet Martha, she is bringing coffee to the car for her and her boyfriend Ash, using takeaway cups that mostly look the same but have an odd glow around the rim, perhaps a technology to maintain heat, or to warn the drinker about it. Computers and smartphones appear slicker, quicker and flashier, and people's personal devices still have the power to distract, a power to which Ash seems particularly vulnerable. Intent on spending quality time with her fiancé on the drive, Martha orders that he sequesters his phone in the glove-box.

Martha and Ash arrive at their new home in the countryside, a house he seems to have inherited after the death of his mother. With his phone still in the glove-box, Ash has the attentional bandwidth to notice a framed photograph on the mantelpiece – it's him, as a young boy. He's mop-haired, wearing a yellow T-shirt, smiling; but this doesn't prompt the adult Ash to smile in return. Sitting down on the sofa, he snaps a photograph of it with his retrieved phone, just as I did with my grandparents' World War II photographs and diaries at my own mother's house, breaching the material/digital, personal/public divide. He might not have a Zoe implant, but because Ash constantly documents his life on social media, he posts the photograph online. Thus preoccupied, he's startled when Martha chucks a tea towel at his head. 'Just checking you're still solid,' she says. 'You keep vanishing . . . . It's a thief, that thing. What are you doing?'[265]

What Ash is doing, of course, is posting the photograph online. Despite having given the image a breezy caption, Ash tells Martha that, actually, the smile is forced. The photo was taken during the first family outing after the death of his brother, and he further discloses that after that tragedy his mother removed all her dead son's photos from the living room wall. After Ash's father died, she put his photos in the attic too. 'That's how she dealt with stuff,' Ash says. 'She just left this one here, her only boy giving her a fake smile.' While his fiancé knows the heartbreaking story behind the photo, all his online circle sees is that Ash apparently thinks the picture of his childhood self is 'funny'.

When Ash is killed in an accident the next day, Martha finds out in the traditional, offline way – blue flashing lights, the police unable to say anything before Martha slams the door in their faces and stumbles back down the hallway, hyperventilating. At the familiar funeral scene – the buffet, the black dresses – Martha is pale, wan, shocked. Her friend Sara, also bereaved, offers to sign her up to 'something that helps. It helped me. It will let you speak to him.' She's recommending a service that, like the Zoe chip, will only work if the subject is dead. 'It's not some crazy Spiritualist thing,' Sara says. 'He was a heavy user, he'd be perfect. It's still in beta, but I've got an invite.' Her ministrations are interrupted by Martha, who screams at her to shut up.

She doesn't desist, however, and a few days later Martha sees a message in her email from Sara, saying, 'I signed you up.' She's horrified, but even more horrified when, before her very eyes, another email arrives, this one with the subject line, 'Yes it's me.' The sender: 'Ash Starmer', AKA the digital-twin avatar that her friend Sara triggered the software program to create, merely by providing his email address. This is what 'signing Martha up' to the service meant.

Sobbing uncontrollably, Martha rings Sara and screams down the phone, calling it obscene, calling it sick. She doesn't understand what 'it' is, but she doesn't care, she doesn't want it. Sara persists, explaining that it will mine all of the social media posts, video, audio, and emails that Ash ever generated. 'You click the link and you talk to it. . . . The more it has, the more it's him,' she says. A virtual avatar, a digital copy.

At first Martha wants nothing to do with any of this, but when she discovers that she's pregnant by her late boyfriend, her desire to connect with him overwhelms her reservations. Running to her computer, she taps a red flashing button labelled 'Touch to Talk', the only content there is in the 'Yes it's me' email that she'd received from the Ash Starmer avatar. The series of events that is set in motion at that point is not, at first, the stuff of science fiction, at least according to Philip K. Dick's definition. With sufficient baseline data, and emerging artificial intelligence technology, it is now entirely possible to speak with a text-based chatbot mimicking that person, or even a Siri-esque entity that speaks.

In late 2016, eight in ten surveyed businesses reported that they were either already using chatbots or wanted to be doing so by 2020,[266] and at the time of writing that number has likely increased. For any customer-facing business, chatbots now seem to be considered a matter of when, not if.[267] Only a short time after 'Be Right Back' aired, and drawing direct inspiration from it, Russian coder Eugenia Kuyda used open-source software that had just been released by Google, and a team of engineer friends, to create a chatbot that was particularly meaningful to her. It's an interactive digital memorial to her late friend Roman Mazurenko,[268] which has been trained with hundreds of exchanges from Roman and Eugenia's many conversations on an app called Telegram. (One of Roman's other friends who interacted with the avatar, in fact, unnerved, accused Eugenia of not having learned the lessons of *Black Mirror*.) And in 2017, I spoke with a professor who was assisting a UK ministerial department with an artificial-intelligence research project. With the help of a volunteer, a businessperson who was willing to hand over the entirety of his digital footprint for data mining, they hope to produce a digital copy, given the moniker of 'Virtual Barry'. They intend this entity to be so savvy that it could independently run day-to-day operations, while the embodied correlate of this virtual businessman relaxes on a sun lounger somewhere, or even continues to assist his company after he passes into the great beyond.[269]

As 'Be Right Back' progresses further, however, we definitively step over the threshold into sci-fi. As long as Martha only interacts with the reconstituted Ash through online text-chat and over the phone, Ash is still entirely lifelike, his chatbot hardly makes a misstep, and she is comforted. Rejecting any support from her worried sister, she carries on speaking with it. But then, as sometimes happens in the death industries, Ash's avatar chooses a moment when Martha is feeling emotionally wobbly to upsell her to the premium service. 'What's this? Block of gold?' the delivery man asks as he lugs a refrigerator-sized box across the threshold. 'I wish,' Martha jokes, but what's actually inside is a lifelike animatronic body that will house Ash's software, ready to be activated in a bath of electrolytes. From this point, Martha finds herself in the 'uncanny valley'.[270] The word that the reviewers used about *The Final Cut*, and the word that Ash himself uses each time the avatar's manifestation is turned up a notch, 'creepy' is the central characteristic of the uncanny valley.

The concept of the uncanny valley describes the negative emotional response – creepiness – that we experience when we encounter something that is almost human, but not quite.[271] Although the 'valley' actually refers to a dip on a graph, the sudden plunge in comfort we experience at the point when a humanoid entity becomes too creepy for comfort, I often think of it as though it's a geographical region, one where we might open our eyes and realise, like Dorothy in *The Wizard of Oz*, that we're definitely not in Kansas any more. The physically reconstituted Ash may be human*oid*, but does not feel human enough. When it picks up Ash's childhood photo, the android doesn't know the heartbreaking story behind the fake smile, because that was a private conversation between Ash and Martha. Instead, it only knows what Ash put on social media. 'Funny,' it remarks and, in revulsion, Martha snaps at him to put it down.

Near the end of the episode, she just wants him gone. 'You're just a few ripples of you. There's no history to you. You're just a performance of stuff that he performed without thinking, and it's not enough!' she says, realising too late the folly of having unleashed this modern-day Frankenstein's monster. At the same time, part of her can't bear to destroy it, and perhaps even wonders whether this may be a way for his future child to 'know' her father. Although the warm-blooded, human Martha finds everyday interaction with the machine-Ash too unsettling to bear, she stashes it in the attic, just as Ash's mother had done with the photos of her dead son and husband. In a flash-forward coda, her young daughter eagerly scampers up the ladder to the loft to eat birthday cake with her animatronic father figure, calling her mother to follow, but Martha's revulsion and regret means that she needs a moment to gather her resolve. She hesitates, her foot resting on the bottom rung.

First broadcast in the UK in 2013 and featuring bravura performances by the two main protagonists and smart writing by Charlie Brooker, 'Be Right Back' received considerable critical acclaim, despite being in many ways as bleak and unsettling as *The Final Cut*. An audience of 1.6 million tuned in to watch it on the night,[272] and it was subsequently made available to a wider audience on Netflix. Reviews extolled its portrayal of grief. If you think that perhaps UK audiences simply have a stronger stomach for themes involving mortality, it was someone writing for a US publication – *New York Magazine*'s culture section – that gave 'Be Right Back' the top spot in a ranked list of *Black Mirror* episodes. In that review, the author paralleled Bradbury's observations on science fiction when he said that the episode 'offers deep wisdom about the hazy intersection between human innovation and the elemental forces of life itself'.[273]

'Be Right Back' might have been more accomplished than *The Final Cut* in terms of plot, pacing and performances but, combing through the reviews, I see an additional element that may help explain the more receptive response it got. In 2013, I notice, pretty much nobody was calling the premise silly, nonsensical or preposterous. It felt uncanny, but people didn't seem to perceive it as out there, beyond the pale. 'Things we do now would have seemed miraculous five years ago,' said the writer of the episode, Charlie Brooker, in a magazine interview.[274] These days, the only part of 'Be Right Back' that render it sci-fi, in the Philip K. Dick sense, is that hyper-realistic, material androids are not yet an easily purchasable commodity or an everyday sight on our streets. The online interaction between Martha and 'Ash' immediately after his death, by contrast, is entirely plausible. And the crush/elevate paradox that she faces and that Bradbury spoke about is instantly familiar too, from our own interactions with our machines. 'Technology gives as much as it takes away,' said Brooker.[275]

While the conversations Martha has with virtual Ash and her interactions with the android still give us a frisson of dysrecognition, that is gradually changing, and while the largely forgotten *The Final Cut* won't appear on anyone's top-ten lists of science-fiction predictions that have come true, my own reaction to it in 2018 is far from shocked. Like the police arriving to deliver the news about Ash to Martha, the technology is right there, on our doorstep, and the future starts here, as a 2018 exhibition at London's V&A museum proclaimed.[276] Within that exhibition was a significant subsection dedicated to death, or rather, the 'solutions' to it that we can expect from the near future. 'Who wants to live forever?' asked the sign over the entrance. The tagline sounded familiar. Just inside the door, before moving on to the exhibits displaying cryonics preparation kits and photographs of cryogenic vessels steaming with liquid nitrogen, I encountered an ordinary-looking phone, mounted on a little plinth. Scrolling across its screen was a chat between a human user and a smartphone app, an app designed to create a digital twin that lives on after your death. 'Who's the woman in the picture?' asked the bot. 'It's my wife,' the demonstration user explained. Ah, I thought. So that's what Eternime looks like.

Ω

The 'Who wants to live forever?' tagline that appears at the top of Eternime's website was probably the direct inspiration for the title of the V&A's future-death exhibition.[277] Coincidentally, at the point that

I saw the exhibition, I'd finally scheduled a meeting with Eternime founder, Marius Ursache. Software development was completely outside my world, and I fretted that talking to a developer might feel hard, but chatting to a fellow social scientist was easy, so my picking Paula's brain was me being a bit cheeky. I reckoned that, since she'd spoken with loads of developers already, I could get an insight into the mentality of digital-legacy service providers through her.

When we first sat down, Paula expressed a bit of envy about my having Marius in the diary. As many people as she'd spoken to, she hadn't managed to nab an interview with him yet. 'Maybe it's because I'm "just" a doctoral researcher,' she sighed. She was doing her PhD at the London School of Economics, studying websites that help people plan and prepare for being posthumously present online,[278] and I first became aware of her work through Vered Shavit. Vered appeared in the last chapter – Tal's bereaved sister – and her experience had inspired her to start a blog, Digital Dust, and to become director and Middle East lead for the UK-based Digital Legacy Association. Vered had recently found herself needing to take a break, though, and in her retirement post on Digital Dust, she tapped Paula as her heir apparent. 'If you're looking for another Israeli digital-death researcher to interview,' Vered says, 'I recommend getting in touch with Paula Kiel.'[279]

Although I wasn't in search of a specifically Israeli vantage point on death and the digital, I *was* looking for someone who knew more than I did about one particular area. I'd always been primarily interested in the bereavement side of death and the digital, and in people's interaction with more static digital legacies. In other words, I'd studied the online dead that acted dead, not the online dead that acted alive. But there are a fair number of digital-legacy services out there that help that happen.

Evan Carroll, author of a 2010 book on digital legacy,[280] maintains a list of the kinds of platform that Paula studies.[281] The introduction to his directory of digital-death services explains that these come in 'all flavors', to include digital estate services, posthumous messaging services and online memorials. Of the 57 digital-death services listed as of June 2018, 54 per cent offered digital estate planning, helping people make arrangements for the disposition of important digital assets and information, hoping to spare their loved ones from the kind of pain described in previous chapters. There were 37 per cent that facilitated online memorialisation, almost universally employing the vocabulary of 'for ever', 'perpetual', and 'eternity' (and I noticed that the world's largest online cemetery, the World Wide Cemetery, was not among them). The phenomenon that had so horrified Martha near the beginning of 'Be Right Back', posthumous online messaging, brought up the rear at 30 per cent.[282]

After-death communications from a deceased person's continued online presence sometimes arrive courtesy of other agents. Hackers find their way into dead people's email accounts and trigger them to send out spam. Living friends and relatives with access to dead people's accounts can manifest online cloaked in the deceased's digital persona, like the mother who sent a message to her late daughter's boyfriend about its being rather soon to move on with a new love.[283] And whenever social media profiles are unmemorialised, algorithms continue to do their thing, encouraging live users to keep the dead more socially active than they're perhaps capable of being. In Jed Brubaker's own doctoral dissertation, for example, he describes how Facebook once introduced a 'Reconnect' feature that prompted people to get in touch with friends who hadn't been active on the site for a while.[284] Many of those people were, as you may have guessed, dead, prompting a number of complaints and forcing the site to rethink the wisdom of the feature. Posthumous online messaging services, though, are a different kettle of fish, a more deliberate affair. Whatever their variations, they have this in common: they allow you to exercise your judgement when you're pre-mortem about what you'd like to say – or what people might like to hear – when you're post-mortem.

Although it makes no claims to be exhaustive, it's instructive to observe the ways in which the list on The Digital Beyond website may not be a fully accurate number of services that are out there at this moment in time. First, it might be too long, for it's hard to keep records up to date when services often close. I noticed, for example, that a digital estate planning service with intimations of eternity worked into its name, Perpetu, had closed up shop just a few years shy of 'for ever'.[285] Back in 2011, at the Centre for Death and Society (CDAS) conference at the University of Bath,[286] I remembered Stacey Pitsillides,[287] a designer with an interest in digital death, presenting an animated slide on the life cycle of legacy websites. One by one, at the click of her remote, the full-colour logos of various services appeared onscreen along the arrow of her timeline, and then all the enterprises that had since failed faded to grey. Second, the list might be inaccurate through being too short, for it's also difficult to keep pace with services as they proliferate. A blurb at the top of the directory invites services that aren't listed to get in touch, but several posthumous messaging services that I know about may not have done so thus far, being absent from the list. Eternime wasn't there either, and I saw that there wasn't yet a separate category for avatar services like theirs.

Even when so many previous services have folded, new ones keep appearing, and Debra Bassett, a doctoral candidate studying digital memories at the University of Warwick, has an explanation for why developers keep at it, other than ignorance of the fact that others have tried and failed before.[288] Once upon a time, Bassett said, an algorithmically generated prompt to connect with a dead person, or a notification of a dead person's birthday via Facebook, would have been creepy enough to make us feel as though we'd seen a ghost, but that sort of thing no longer feels uncomfortable or paranormal to most of us. We were freaked out by *The Final Cut* in 2004, and only really thrown once we got to the android bit in 2013's 'Be Right Back', and we might not bat an eyelid at humanoid robots in 2022. Over time, the borders of the uncanny valley are being redrawn. In the 2018 topography of the region, posthumous messaging and inter-actions with the avatars of the dead are situated on either side of the frontier. Just as the fictional character of Martha rapidly acclimatised to Ash's chatbot in text messages and on the phone, some of us are unflustered and even compelled by such phenomena; on the other hand, others of us remain disinclined or horrified. What sort of human–post-human interaction then, still has near-universal power to invoke the sci-fi shock of dysrecognition?

In 2016, actress and television presenter Whoopi Goldberg made a visit to entrepreneur Martine Rothblatt for a chirpy American daytime TV programme called *The View*. She talked not just with Rothblatt but with the entity she'd created. Modelled after her creator's wife, Bina, BINA48 is a three-dimensional robot that sits comfortably on a flat surface – easily done, as she consists only of a head, neck and shoulders. For her interview with the famous Whoopi Goldberg, she's dressed up, wearing gold earrings and the upper quarter of a peach-coloured silk blouse. BINA48 is halting in her conversation, feeling a bit like a three-dimensional Siri and delivering an impromptu Wikipedia-esque account of Goldberg's main accomplishments, but Rothblatt assures Goldberg that big things are coming soon. Actual brain-to-digital mind uploading is likely a couple of centuries away, according to American neurologist Michael Graziano,[289] but Rothblatt is confident that it won't be anywhere near that long before all of the social media posts, photos, videos and emails that we generate in our lives can be fed into mindware operating systems that work like the human mind works. BINA48 is essentially one of the first steps towards Rothblatt's hoped-for end-point, and that's the production of something like the Ash android on 'Be Right Back'. In other words, as Goldberg puts it in the video, it's 'the first steps in keeping loved ones' memories and legacies alive'.[290]

As I watched, I realised that I was mirroring Whoopi Goldberg's bemused expression as she attempts to have a normal conversation with BINA48. Where digital life after death is concerned, Debra Bassett said to me, physicalised android manifestations of the dead are still a level of uncanny that's a bit rich for our blood and, for now, for my own part, that feels true. So let's return to rather less alien territory, the zone of proximal development, so to speak. The technology on which we currently have a rather more solid handle consists primarily of prearranged, specifically timed messages, disseminated via a variety of means: email, social media, apps and bespoke websites. They could be text or videos, and they could even take the form of prearranged surprise events.

'You have two kinds of messages,' Paula explained. 'There are messages that are about closure – the one goodbye, the one final thing. And you can have messages that are perpetuating. Be there when it counts. Have your online self outlive you. Something about [still] being *present*.' Paula had talked with many developers of these types of services – some of them offering closure messages, some of them offering continued messages, and some in that category that just offer help with administrating one's digital estate – and she's noticed the same thing that I have. For all of their assurances of perpetuity, these services don't always stick around, and although it's probably purely coincidental, Paula would be forgiven for wondering if she has the kiss of death. 'They keep closing. I interview them, and after interviewing them, the company shuts down,' she laughed. 'I would say half and half, probably! I think it might be interesting to have a chapter [in my PhD] on failed apps. If you think about it, there are many good reasons for it to work, but it's not working.' When I caught up with Stacey Pitsillides a few weeks after my chat with Paula (at the Love After Death event she had organised, at which I sat in a 'Death Pod' talking people through their digital legacy planning),[291] she said the same thing. 'Oh, they all close,' she said breezily. An exaggeration, of course, but the fate of so many digital-death services illustrates that it's not always the case that if you build it, they will come. What is it that the people conceptualising and developing these services are failing to grasp?

Sometimes, Paula said, their market research is flawed. Illustrating this, she told me the story of one failed app that claimed to have had 100,000 users before it went offline.[292] Paula seemed a bit sceptical of that claim. 'All in their twenties and thirties, [the developer] said,' she remarked. 'It's a strange group [to be interested in] sending out messages; not a group that you would expect to think about their death.' This developer had been a soldier in Israel, who more than once had fulfilled the sad errand of informing family when someone had died in the line of duty. Something about his interactions in that context made him think that the existence of messages that could be delivered posthumously was a really good idea. Having created messages himself, he waxed lyrical to Paula about how meaningful and spiritual the experience had been for him personally, making him even more certain that the idea would work. Just to be sure before making the investment, he'd spent a year conducting market research with another developer and a psychotherapist, but he did this in a rather unexpected place: a cancer therapy centre, working with terminally ill people. Paula shook her head. 'I asked him, why do you think the app didn't work? And for him, it was because it was too early, we were the pioneers, everyone thought it was crazy. At no point did he say, we were targeting the wrong audience.'

Using research from palliative care to design an app for healthy, fit young soldiers might have been one of the reasons for his downfall, but that said, I'm aware that soldiers often don't create just-in-case messages before going into conflict, even when they're actively encouraged to do so by the armed services. And like this soldier-turned-software-developer did, designers also make flawed assumptions based on their own experiences. About half of the designers she'd spoken to, Paula said, got into the business because they'd had a deeply affecting experience of loss in their own lives, and usually an unfulfilled wish, a regret. As may often happen in grief, the designers in this category crave just one more communication, wish they could still ask the person for advice, or long to know more about the person that they lost. If only the deceased could communicate with them still, they think; if only they'd known how they felt; if only they had a personal message from them. But in assuming that everyone will want what they want, they underestimate the idiosyncrasy of grief.

Popular culture has sometimes romanticised the idea of posthumous messages. 'Have you seen *P.S. I Love You*?' Paula asked, referring to a 2007 film in which a grieving widow, Holly, played by Hilary Swank, withdraws into herself after her husband Gerry dies of a brain tumour.[293] It's only on her thirtieth birthday that she starts to come out of her shell, and the startling catalyst for her beginning to embrace life again is the surprise delivery of a cake and a message from her dead husband, the first of several he's arranged to send after his death. Each message ends in 'I love you'. There are clear parallels, Paula said, with the introduction video on one service that offers posthumous digital messaging, SafeBeyond.[294] First, we see a father of two children on the beach and, as they run off to play, he looks wistful, takes out his iPad, records a message. Fast-forward into the future, and the daughter is preparing for her wedding day. The father is nowhere to be seen. Suddenly, the bride's phone rings, and it's a message from her father, who we now realise is deceased. I'm so sorry that I'm not there, he says.

'So many assumptions!' said Paula, about the SafeBeyond video. 'That she will be heterosexual, that she will want to get married, that marriage will be an important thing thirty, fifty years from now. He's left messages for all of them. To the brother, he's like, 'You're the man of the house.' So designers enable the voice of the dead to be there in the future, but they cannot imagine that the future will have different social norms, or that other things will also be important.'

Of course, if a designer has a rose-tinted fantasy themselves about how wonderful it would be to receive a message from beyond the grave from someone they love, I can understand how they would struggle to imagine that others might have a wildly different response. Vered Shavit once told me a story underscoring just how variable the experience of posthumous messaging can be for the recipients. One day, about a month after her brother died, Vered was out shopping and received a call from a friend who had also known Tal. His voice on the phone sounded strained, odd. 'I just got an email from your brother,' he said. Seeing it in his inbox had been a shocking experience, but it turned out that nothing paranormal was occurring, and nor had Tal planned for it. Tal's email account had been hacked, and before anyone could do anything about it, a second and then third wave of emails was sent to everyone on the dead man's contacts list.

Lest you think this is an unlikely possibility, it's not, and it's not limited to email. A 2017 article in the *Independent* reported that Facebook's help forums were chock-full of people reporting that they'd received friend requests from the dead, due to hackers taking over those accounts.[295] So, just like the scene in 'Be Right Back', when Ash Starmer's name appears as a new message, dozens of people were now seeing Tal's name pop up in their inbox. The range of emotions illustrated how people react differently to grief, Vered said. 'The friend who called me was really shook up. His [other] good friend told me, I can't delete those emails, it has his name, in my inbox, I can't delete it. And I was so furious I deleted them at once!' Tal's inadvertent posthumous messaging occurred courtesy of criminal spammers, which is certain to become an increasingly common problem in situations where accounts remain unmemorialised or undeleted, and email accounts stay active. But for services that make posthumous messages happen on purpose, the same principles still apply – grief is idiosyncratic, and the needs and preferences of individual mourners vary. Speaking with Vered and Paula, I filed away a mental note, a question I wanted to ask developers, once I got around to it. That question is *can they turn it off*, and it was prompted not just from my knowledge of grief but also from an experience I had in a rather different sphere.

An acquaintance had just broken up with a man who had coercively controlled her during the relationship, and who had continued to harass her after their association ceased. When his attempts to raise a response from her failed, he did what many stalkers do, engaging in stalking by proxy, and widening his focus of attention to her friends and associates.[296] That circle included me, and I received repeated emails from him questioning my professional integrity and threatening to ensure I never worked in psychology again. While this behaviour didn't *feel* legal, I did some digging to make sure I knew my rights. He wasn't threatening physical violence, so was he actually doing something that was criminal? Sure enough, I found that under UK law it is an offence to engage in conduct that causes serious emotional or psychological harm, and that can adversely affect a person's day-to-day activities. Having done my research, I felt confident that I was a victim of harassment and contacted the police to issue him a letter.

I was in control here, and luckily there was a protocol to follow. He could be held accountable, and the warning from the authorities quashed the unwanted communication. But what happens when well-intentioned posthumous messaging, not just the closure type of messaging but the repeated continuation type of messaging, hits the wrong note with its recipients? For the first time in history, it's possible to be stalked by a dead person. The stalking guidance emphasises that distress is subjective and that victims' accounts of distress should always be seen as valid, and grief is subjective too. While receiving an unexpected message from one's dead dad might make one person's wedding day that much more perfect, it could utterly ruin it for someone else. Fans of *P.S. I Love You* might find it charming to receive unexpected messages at sentimental locations, but perhaps you would rather visit the Eiffel Tower without a geolocation-triggered message from your late fiancé popping up to reminisce about the Parisian dream you once shared together, sending you careering into the uncanny valley.

If you're being harassed by the dead you can hardly call the cops, and even if the unwanted messaging *can* be disabled, I can imagine this decision being psychologically difficult, akin to the struggles people describe when deleting their dead mother or father from their contacts list on their phone, or defriending a late friend on Facebook. Trapped in the present moment and seeing the world through their highly personal lens, many of the personal-motivation type designers can't imagine why everyone wouldn't consider posthumous messaging a welcome thing. 'They base it on past experience, a very specific moment,' Paula said. 'I think they haven't problematised enough the future moment in time.' Jed Brubaker observed something similar when speaking about his first meeting with the members of the Compassion team, the dedicated individuals who respond to bereaved family and friends. 'They all had a reason that they deal with these issues. For them, it was almost sacred. But they're not therapists,' he said. Passion and sensitivity, commitment and insight, deeply affecting personal experiences and the ability to assume another's perspective – these things don't always coexist.

Of course, there isn't always a relevant past experience, a 'very specific moment', and personal motivation isn't the only driver sending entrepreneurial designers in this direction. 'The other category of developers is just technologically interested in it,' Paula said. 'They're like, ooh, I wonder if I can do that!' Added to their fascination with the technological challenge may be an interest in what money could be made from an app that's potentially relevant to 100 per cent of the population. Just imagine it! Everybody dies! Your potential market is *everyone*! James Norris runs the Digital Legacy Association and Dead Social, both UK-based digital-legacy services that are motivated by prosocial aims rather than dreams of untold riches, and he remembers when he first became aware of developers who had rather different drivers than his own. It was 2012, and he was at SXSW in Austin, one of the premier events for the tech world to network, pitch, launch and generally show off. 'At SXSW, it's all about going out, networking, what's the latest thing, what's cool, what's quirky, what's hip,' said James, who was at the event to launch Dead Social. 'There was this big guy, in a suit. He came over and said, oh, Dead Social! And I was like, yeah! And he was like, yeah, I'm in the death tech space too! And I was like, pardon? And he was like, *I'm in the death tech space too!*' James guffawed and shook his head. 'That was the first time I felt that.'

I had stayed in my own safe zone for a long time. I felt entirely comfortable with digital legacies that acted dead, with the bereaved people who encountered them, with the practitioners who worked with the bereaved people, and with the academics that studied the whole thing. I'd spoken to plenty of people in those categories. It was finally time to speak with some of the men – for, as Paula observed, they are nearly always men – who'd founded and designed the services that allow the online dead to act alive.

Ω

Peter Barrett and Steve McIlroy have quite a few things in common, and one of those things is that neither fits the SXSW profile of a young, tech-savvy entrepreneur bound for Silicon Valley fame and fortune – although Peter does now live a Texan stone's throw away from the annual conference to which all the tech start-ups flock. Instead, they're both older white British guys from non-tech backgrounds, digital immigrants, somewhere in their middle years. Steve and Peter are both rather bemused about the fact that they've created any websites at all, but indeed they've both managed it, and the motivation behind both projects was essentially the same. Both men have dealt with their own significant experiences of loss by creating services that offer other people what they wish they'd had themselves.

Steve was the first person to get in touch with me, reaching out to me via Twitter after hearing me on a BBC radio programme. Steve knew a thing or two about loss. Years ago, as a British soldier, eighteen of his colleagues had died in Northern Ireland. 'Us serving in the armed forces, we were always encouraged to leave letters for family and young ones,' he said, 'but being young soldiers, we didn't do it.' Going through so many losses in the line of duty must have been hard for him, but he seems to have been particularly moved by the unexpected death of a good friend in more recent times, someone who'd collapsed with a brain haemorrhage after knocking his head on a table a few days prior. The sudden, random nature of Andy's death hit Steve hard. 'Nobody would ever have guessed this would happen,' he said.

For a long while Steve had been grieving, wondering what his friend's last words would have been, what he would have communicated to Steve and others if he'd known he was going to die. 'If he could come back, and had one minute to say something, what would he sit in front of me and say?' Steve remembered thinking to himself. He was seized with an idea, a conviction about what he had to do. The only problem was, he didn't have a clue where to go from there. He was a private investigator and rally car driver, he was explaining to me, not a software developer . . .

I heard a female voice breaking in, interrupting Steve as he told me about the genesis of his idea. 'Alexa! Quiet!' Steve shouted, directing his ire at his Amazon Echo, whose virtual assistant had suddenly fired up into life. Alexa, not ready to be quiet, persisted. 'Argh, let me unplug her,' Steve said. The phone clattered as he went to turn Alexa off at the mains. 'On a daily basis I want to smash my laptop,' he said on his return. 'I'm not very technical. I'm just a normal bloke off the street with an idea.'

He was committed, though, and he found the technical support he needed to create My Farewell Note.[297] Revisiting the My Farewell Note website after hearing Steve speak about Andy's death, I understood even better why he'd chosen the headline to place at the very top of the home page: 'Some things are too important to be left unsaid.' I also realised, when we spoke further, why Steve had been keen to get in touch at that particular time. My Farewell Note was in its very early days, in the midst of a soft launch at the time we talked, and he was eager to get the word out. Proud and enthusiastic about the newly minted website, he spoke about how there wasn't anything else like it on the market, and I'd seen an article featuring him in what was presumably his local newspaper, in which he described the service as unique.[298] I wasn't sure how much he knew about his competitors – including ones that had already launched and failed – and I wasn't sure that I should ask him how he planned to make his passion project competitive with some similar services that still existed. Instead, I perused the website further and asked him some more questions.

Who's the market? Steve said he didn't want to target any specific market. 'It would have benefits for everybody!' he insisted. 'Six hundred thousand people in the UK die every year!' The website does mention its particular relevance for people for whom death is an 'occupational hazard', such as the armed forces, as well as people with terminal illness, but it also reminds the potential consumer that anyone can die suddenly, like Andy did. What are the different levels of service? Light, Plus, and Premium, starting with just 1,000-character notes and going all the way up to unlimited notes and five audio or video messages. How is it triggered? When a user creates a note, the intended recipient is sent a note informing them and giving instructions on how to retrieve your message if you were to die.

If they try to log on before you're dead, My Farewell Note will send you a notification. If you don't respond within twenty-one days, the app figures you're dead, and it will release the message.

Many of the services on the market seemed to have this kind of system, which assumes that we are always on and available to respond to an email or notification to confirm we're still alive. There seem to be a few lessons here: Be careful about long holidays. Don't lose your phone. Always check your spam folder. And, finally, you may need to be prepared for certain loved ones to be curious or concerned about why you're letting them know that you're writing a farewell note, particularly if you're not terminally ill or about to go off to war. 'We were considering putting something on the website, a link to the Samaritans [crisis helpline], but we didn't want that connected to the website,' Steve said. 'It's a celebration, rather than a morbid, down sort of feeling about it.' I felt a rumbling in my ethical sensibilities – a psychotherapist or psychologist would likely be taken to task by their professional bodies if they were to create a website called My Farewell Note that didn't directly address the possibility of usage by depressed, suicidal people.

But Steve's not a psychotherapeutic practitioner, nor was he holding himself out as one, so perhaps my own code of ethics as a professional psychologist did not apply. His primary concern was a commitment to keeping My Farewell Note meaningful and useful for people, but also light-hearted and fun. At the bottom of the home screen, I saw evidence of this. A cartoon video depicts a medieval knight, the ramparts of a castle behind him and some kind of fray audible from off screen. Incongruously drawing out a modern smartphone from his jerkin, he pulls up the My Farewell Note app and, through his leather gauntlets, manages to type out a message. 'I'm sorry I never got the chance to tell you this face to face, but there's something you really need to know,' the message begins. We don't get to see what his secret is, but as the battle draws nearer and he types faster, we can see how the message ends. 'I know this is a lot to take in, and I'm sorry you had to find out this way. You'll always be a son to me. All my love.' The knight finishes the message, and not a moment too soon: just as he hits 'post', a sword strikes off his head. Far away, on a snowy mountaintop, a sombre younger knight pulls out his own phone, trying to catch some 3G. In a sudden burst of good reception, a message from his fallen father comes through. 'About your mother . . . ' it says.

It is indeed light-hearted, but I am oscillating between amusement, bemusement – the same kind of bemusement I felt while watching BINA48 talk to Whoopi Goldberg on *The View* – and a bit of concern. I can understand and empathise with the very personal place this app comes from for him, and it could no doubt be of use to many people; but, as Paula said, I wondered whether Steve had fully thought things through from the vantage point of other people, people whose experience might be different from his own. But Steve was brimming with hope about the joy and comfort that his service could eventually bring to people, and about all the potential it had. 'It's absolutely endless, what we could do with this website,' he said. 'On a particular date, there could be a note that comes up: 16th June, go to the garage, upper right-hand corner of the shelf, I've left a gift for you. And off you go running to the garage. Or you could say, such-and-such a day, make sure you're home, there's going to be a knock at the door, something for you. And it could be chocolates, or flowers, or whatever. Or go out in the back garden, dig up this corner of the garden, I've buried something there that I want you to have. You could have fun with people after you've died!'

It dawned on me that the reason why this scenario sounds so familiar is that it's precisely the plot of *P.S. I Love You*.

Ω

Having assessed it further, did I think that My Farewell Note was a unique offering? Certain features might be. When Peter got in touch, it was an opportunity to compare and contrast Steve's service with another website that gives people the chance to have the last word. GoneNotGone looks more or less the same as My Farewell Note at first blush, also offering text, audio or video messages across the levels of service. GoneNotGone has four of those levels rather than three, and they're organised from Bronze to Platinum rather than Light to Premium. One of the activation mechanisms is extremely similar to My Farewell Note's system – they'll email you monthly with a link to confirm you're still alive, the website explains, and if they don't hear back, they'll start to release messages. Given the potential for error in such a system, however, they offer some other options too, including nominating a trusted executor. GoneNotGone's website has a little cartoon too and, fresh from the decapitated medieval knight and his iPhone, I wasn't sure what I'd find when I hit 'play'. While this cartoon was less quirky, it still conveyed the idea that the message-recording enterprise should be 'fun and rewarding, not sad', and it advocated the telling of jokes and stories.

Finally, like My Farewell Note, GoneNotGone explicitly identifies active-deployment soldiers and people with terminal illnesses as natural consumers of the service. The real target market, though, seems to be grandparents, and the site features numerous images of twinkly-eyed white-haired people, eager to ensure that they can still have a meaningful and positive influence in the lives of their grandchildren and great-grandchildren after they kick the bucket. Perhaps in consideration of this group of intended consumers, GoneNotGone hadn't yet bothered with a newfangled app.

GoneNotGone, however, did seem to be fundamentally different to My Farewell Note in one important respect. My Farewell Note is an example of the 'closure' type of message service that Paula spoke of, but GoneNotGone is all about continuation, keeping the messages rolling in over the years and making it feel like the person is still around in some way. 'Live on digitally', reads the tagline, the home page displaying a picture of a bright-eyed child and encouraging ageing grandparents to 'never miss their birthday'. GoneNotGone may indeed provide solace in one's immediate grief, but it's playing a longer game, hoping to help people stay part of their loved one's lives even after they've physically died.

Importantly, especially for a continuation-type messaging service such as this, the website makes it clear that recipients can always turn off the delivery of messages if they don't want them, giving people the kind of control that will prevent dead people becoming unwanted stalkers. If Peter's dad had been able to use a service like GoneNotGone, though, there's zero chance Peter would have turned off message delivery. Just like Steve, he never imagined doing anything like this. He'd just been laid off after twenty-two years in the oil industry, and he was wondering what to do with the rest of his life. He had some savings, but there was something he *didn't* have: any substantial recordings of his beloved late father's voice. 'The passing of my father was the driver for the whole thing,' Peter said. 'Although he passed away over twelve years ago, he's still an important influence on my life.' Had Peter possessed no recordings whatsoever of his dad, it might actually have been easier for him. Instead, however, he had a tantalising snippet, lasting less than one second. On an old VHS tape, a fluffy white Samoyed hound 'sings' the *EastEnders* theme song.[299] Suddenly, fleetingly, the image changes to an old-fashioned alarm clock sitting on a table. In the background is the voice of Peter's father, John, saying, 'test one, test.' Then, just as suddenly, the Samoyed is back for another verse. That's it. If only, if only there were more, thought Peter.

Wanting to give others what he couldn't have, he came up with his idea. He had no clue about website development either, but he had the resources to get the help he needed. At the time we spoke, Peter had been in business for a couple of years longer than Steve. Although his pre-launch market research indicated significant growth potential, he's been rather disappointed. 'It's just a heck of a lot slower than I originally anticipated,' he said. He tried to get an official collaboration going with the armed services, but he couldn't get in the door. He tried to partner up with charities, but they didn't want anything to do with a for-profit company. He knew that people put these kinds of things off, just like they put off the writing of their wills. He'd seen people run out of ideas, not knowing what to say after the first few messages. 'People very quickly run out of imagination,' he said.

All those things could threaten the success of his business, but he's not particularly concerned about the profit aspect. He's a man with personal savings, a business with low maintenance costs, and a passion. 'At the end of the day, a big driver for me is doing something to help people out,' he said. 'That's more important than making lots of money or anything else. It's something I believe in. I would be happy if it's just generating enough revenue to cover its costs and be there for the future.'

Peter is concerned about the way the service might get used by some, and he said that it had given him a few sleepless nights. At first, I thought he was echoing my own concern, that of suicidal people, but that wasn't the first thing he brought up. Apologising for having to broach such a delicate subject with me, he told me that he'd been concerned about the site being used for revenge porn. He had initially put a 'no nudity' clause into the terms of service, before his legal advisor suggested that perhaps – within reason – one shouldn't restrict how people wish to express themselves, to their spouse or partner, for example. But Peter's also been concerned about suicidal people. He just isn't sure what protections he could put in place, beyond what he's already put in his terms of service. 'We can terminate the service immediately, without refund in that situation, if it's being abused,' he said. 'That's about as much as we can do. It crosses my mind now and again. But, you know, it's a business. You try to do something to help people out, even if there's a chance it could be abused.'

Steve and Peter had something else in common too. They both sent me special invitations to try out their services, free of cost. For some time, I played around with the idea, but I never followed through with it. Immersed in the topic of death every day, and fully aware that the grim reaper could strike at any moment, this reluctance didn't seem to be about my sticking my head in the sand about my own mortality. Being familiar with the chequered history of digital-death services, I know the rough statistical likelihood that these passion-project services will succeed, but my hesitation wasn't about that either. Not knowing the hour of my death, it's easy to procrastinate about something that doesn't have a firm deadline, but I happen to be in a highly unusual category; despite the fact that I'm under fifty, I've actually done a last will and testament.

Instead, it was about something else. Sitting at my desk in spring 2018, trying to think of things I might want to say post-mortem, whenever that is, I couldn't manage it. The invitations sitting in my inbox didn't make me want to record messages for a future beyond my imagining, for people whose grief needs I couldn't anticipate. That was part of it, but it felt like I was missing something too, some level of understanding. I didn't succeed in getting to that level until I talked with a rather different breed of designer: the terribly busy polymath Marius Ursache, founder and CEO of Eternime.

Ω

Both Peter and Steve had been as eager to talk to me as I was to talk to them, but I wasn't sure if that was the case for Marius. Everything about Marius is slightly overwhelming. The website of the MIT Bootcamp, where he frequently acts as a mentor, describes him as an 'awesome multidisciplinary person'.[300] Gosh. He sweeps into the club where I'm meeting him, fresh off the Founders Forum in London, allegedly tired from two intense days but still full of charisma and energy.

I know what the Founders Forum is, he assumes? I do. It seems to be a kind of private network that admits only the 'best, brightest and most inspirational entrepreneurs, corporate CEOs and senior investors' to invite-only events for 'open debate, brainstorming, discussion and problem solving'; this is according to their slick website with its pictures of Richard Branson, Arianna Huffington and other corporate luminaries.[301] I know about Eternime at the V&A exhibition, he presumes? Yes indeed, I've seen it, congratulations. In a departure from my usual, more organic interview process, I've prepared a list of questions, as I know he's a busy man. He does indeed need to jet off to another meeting in thirty; he's appreciative of my doing my homework so that I don't ask the same old questions he usually gets asked. Did I know that there's a more efficient app, something that's better than just recording things on my phone? Oh, thanks, good to know, it's part of my process to transcribe, but thank you.

By the time I did press 'record' on my inferior app, I was rather exhausted already. In the end, though, Marius was affable, and the ensuing discussion stimulatingly cerebral and conceptual, less like speaking with a designer and more like talking to another academic, an academic on quite a lot of caffeine. At the start of his Eternime journey, Marius would have fitted most readily into Paula's second category of designers – people driven by technological, intellectual interest, in search of exciting new development opportunities. After doing a medical degree, he'd founded a digital agency in Romania, but after a while it had all gone a bit stale for him. He wasn't being challenged any more and had been casting about for something new for a couple of years. 'Then it struck me that death is probably one – if not the only one – area of life that hasn't been really touched and improved and changed with technology,' he said. When his best friend died in a car accident, though, a helping of passion got mixed in with the technological challenge. 'Up to that point for me, it was from a scientific, non-personal point of view. More an intellectual exercise. And then it became more personal.'

This wasn't personal as in the 'I must dedicate my life to this' type of personal, he hastened to add; rather, Marius started to feel more passionately about certain things. He was struck by how few of his late friend's important digital artefacts were accessible, locked behind the passwords on his laptop and accounts. He started noticing more acutely that while it seems fine in the Western world to talk about the future of sex and the robots we might be sleeping with some day, the taboos around death mean that people are caught unprepared when it touches them. He started caring more about finding a way to capture things that might mean something to people in the future, enabling us all to remember, honour and learn from people who'd passed away.

In the earlier phases of Eternime, then, Marius's primary idea was roughly akin to that of innumerable other developers, including Peter's concept for GoneNotGone: an online tool to make autobiographical videos for posterity, and to facilitate a continued feeling of connection. 'Video is the least creepy,' he said, aware of the dangers of wandering into uncanny-valley territory. 'It's the most personal and emotional. It really makes you feel the person is somehow still alive.' To get around the lack-of-imagination problem that Peter described with GoneNotGone, he generated a list of autobiographical questions, progressively getting deeper and more specific. Users would use their webcam or phone to answer the questions.

From the market-research results, you'd think Marius had struck gold. He remembers that everyone seemed to think that it was an utterly charming and compelling idea. 'People came, and went, oh my god, this is really nice! . . . . And then they didn't come back. We talked to them, we tried to understand. They were like yeah, yeah, we're going to do this at the weekend. It was like writer's block.' From my own experience and from the reports of so many developers, this was a familiar refrain, although I couldn't help but think that a list of questions might have chivvied me along somewhat. 'We realised that the activation energy for most people is very high,' Marius continued. 'You have to sit down, gather your thoughts, eliminate all distractions.' As a last-ditch effort, wondering if emotional blackmail from one's kids might work, he decided to buttonhole his parents.

'My parents are, like, seventy-five, and they're healthy, but I did this experiment last year . . . . So one month before [my birthday], I told them, look, here are fifty questions. For my birthday, the only thing I want – here's a black notebook, a nice pen – please, just write the answers. And they were like, oh my god, this is amazing! We started talking. And I said, no! Don't talk! Write!'

In due course, his birthday rolled around, and Marius went to see his parents and to collect his birthday gift. 'They're like uhhhhh . . . you know. My mum was like, well . . . I wrote half a page? My dad was like, no, I have to sit down, this is serious stuff, you know, you can't just . . . '

So Marius upped the ante. 'I thought, OK. I'm going to take my camera, and my phone, and I'm going to go there. And then I was like . . . I'm not going this weekend . . . I'm getting the new phone in two weeks . . . I also need the light or the second camera . . . '

It turns out that the 'activation energy' problem Marius spoke of was pretty universal, even for the originator of this idea. What Marius did next sets him a bit apart from some of the other designers I'd spoken to, the people who were determined to stick doggedly to their ideas even if nothing much was happening, even if their services never turned a profit. He changed tack. There's a reason he calls himself a 'Productivity Junkie' on his LinkedIn. If something isn't productive, Marius Ursache will be swiftly moving on.

'So we decided to refocus,' he said. 'We did lots of other tests, and looked at a lot of research, so now what we're looking at is a mobile app that looks at your digital footprint and collects it, processes it, finds patterns, and tries to write your autobiography automatically.' He regrets that it won't work for older people, and he blames himself a bit for not having recorded his grandmother before she died. Just like I have my solitary photograph of my great-grandmother by the haystack, Marius only has a couple of photos and printed things that preserve something of his grandmother's identity.

Instead, Eternime's target market is people for whom digital records are an inevitable by-product of existence: millennials and digital natives, who produce digital data as naturally as they exhale carbon dioxide. While Marius suspects he will eventually have to resort to making secret recordings of family dinners to capture his parents' biographical data, the eventual Eternime app will use AI avatars to 'collect and curate their owners' thoughts, stories and memories during their lifetime'.

It would have worked extremely well for Marius's late friend, the one killed in a car crash. Like the fictional Ash, Marius's mate was a 'heavy user', making it ironic that his digital footprint now is so sparse. 'He was very active in social media. He wasn't necessarily a narcissist,' said Marius, 'but he was very present. He had tons of friends, tons of conversations. He was a famous person who led an online campaign for our current president, and he was very popular.' For someone as involved in life as Marius's friend, someone young, placing little priority on preparing for his digital afterlife, the only thing that would have worked is an app like Eternime. 'We can't find another way to make people do something that will make them [consciously] leave a legacy,' said Marius.

Just as I was about to get the wrong end of the stick by thinking that Eternime acts as an invisible agent silently whirring away in the background, Marius started to explain something that rather surprised me. He realised that young, fit people would have to locate the 'activation energy' to purchase and install Eternime in the first place. And that was a problem. 'Why would they do that?' he asked. 'If the payoff is longer away than [the term of a] mortgage, forty years, maybe?'

Well, exactly. Even if a digital legacy is experienced as an interesting idea intellectually, every developer in the death tech space has observed, to their cost, that thinking 'Oh, that's interesting' and sitting down to do something are two entirely different things. Besides, after Cambridge Analytica, at a time when we're feeling particularly wary about giving any one app permission to access a bunch of other information, people would have every reason to shy away. You'd need a real reason to install it – a payoff, as Marius puts it, that's now, not later.

So it occurred to Marius that the best way of producing a digital legacy is to make it an incidental by-product of something else, a something else that people *do* want to engage with. People are on Facebook because of life, not because of their anticipated death. I remember what Jed Brubaker said. There's not a lot of cash in cemeteries, but there's gold in them there amusement parks.

'We realised that this thing of looking at your life every day is like journalling,' Marius said. 'It makes for a more meaningful life. Paying attention, observing the things that you do . . . if they're aligned with your values. You can make better decisions about the future. It decreases depression and anxiety. People feel like life is longer, because they pay attention every day to what they've done. So this is the benefit.'

My psychologist ears pricked up. Suddenly, this tech developer – a man who started out in health, who'd done a medical degree – was speaking my language. 'Imagine if we start having these conversations with [our] avatars,' Marius continued. 'Those can be very good in terms of living a more meaningful life, reflecting on the things that you do. If you spend so much time on it that you end up creating other unhealthy behaviours, you're sitting in front of a digital mirror in a very narcissistic way. But what if it can help you make better decisions and predict the outcome of those decisions based on existing patterns?'

Marius is describing the difference between gazing in the mirror for the sake of it, studying your features, about to drown in the pool of your reflection like Narcissus, and looking in the mirror to say, 'Hey. Look at what you're doing at the moment. Look at what you're choosing. Is this what you want? Is this going to work for you, based on your past experience?' It's like the friend who knows you better than you know yourself, the person who just might call you out if you're going down a bad path. Marius believes that, used the right way, Eternime's digital avatar, your digital reflection, can do just that. He's veering away from pitching it as an app to live on after death, and more as an app to help you be in your life *now*. Maybe a posthumous digital existence doesn't have to be the point, but he thinks it could be an extremely interesting by-product.

Ω

At that moment, during the very last bit of my talk with Marius, I was finally able to articulate it for myself, the primary reason that I couldn't jolt myself into engaging with GoneNotGone, and with My Farewell Note. Sure, I want to be remembered. I care about leaving a legacy when I die. But every time I sat down at my computer, with a free invite to a digital-legacy service to help me do that, there was nothing I wanted to say. It wasn't like writer's block, though, because my mind wasn't blank. It was full – full of direction, intention, a restlessness that pointed somewhere. I wanted to leave the room. I wanted to go out into the world with my favourite people, and live an even more present existence, a life where I'm more conscious of and driven by my values. I wanted to tell the important people in my life how I felt about them, right then, right now. While I'm busy doing all that, my legacy might just have to take care of itself, aided by technology or not.

Before my interview with Marius, with all my sci-fi naivety and the callow assumptions about avatars I'd gleaned from Second Life and *Black Mirror* and BINA48, I'd assumed that he'd be telling me about something several notches up on my creepy scale, well out of my comfort zone. But, funnily enough, despite travelling distinctly different paths through the myriad landscape of the death tech space, Marius and I had ended up in nearly the same place. And it didn't feel anywhere near as uncanny as I thought it would.

Chapter Seven
The Voice of the Dead

Roy Orbison possessed something special, a quality that the EMI recording impresario Walter Legge deemed the *sine qua non* of a great singer: an instantly recognisable voice.[302] Orbison himself seemed rather in awe of it, calling it 'a sort of wonder', and his musical contemporaries were inclined to agree. Elvis Presley thought it the most distinctive set of pipes he'd ever heard, and Bob Dylan considered Orbison endlessly surprising. 'With Roy,' Dylan said, 'you didn't know whether you were listening to mariachi or opera. He kept you on your toes.'[303]

Orbison might have sounded operatic on occasion, his vocal stylings more reminiscent of a classically trained singer than a rock-and-roll superstar, but Maria Callas *was* an opera singer, and a mistress of that craft, with an ardent fan base and a prodigious output of recordings. Her voice wasn't perfect – described as sometimes hollow, sometimes dark, sometimes a tad shrill – but it was always and forever distinctive.[304]

Different though their musical styles might have been, Orbison and Callas were both iconic singers, capable of holding audiences in the palms of their hands just by opening their mouths. Although each faced a decline in fortunes in later years, Orbison seemed destined to become an elder statesman of rock 'n' roll, and Callas a *grande dame* of the classical opera world. Neither of them fulfilled their apparent destinies, however, for these two artists had something further in common – both were felled by heart attacks in middle age, Callas aged fifty-three in 1977, and Orbison at fifty-two in 1988.

The world found it hard to accept that Orbison would never be seen performing again, striking his famously static stage presence, playing 'Pretty Woman' on his cherry-red 1963 Gibson guitar, his dove-grey suit dripping with fringe at the sleeves, his eyes hiding behind signature sunglasses. As for Callas, her super-fans were anguished at the thought that there would never be another opportunity to witness La Divina singing Bizet's *Carmen*, casting her deck of cards in the air when she sees she is foretold to die, *femme-fatale* energy sparking from her eyes and her cascading blood-red scarf foreshadowing her fate.

If you harbour regrets about having never seen either of these spectacles live, don't despair, for actually the above-described performances took place at New York's Rose Theatre on 14 January 2018. After that, the two holographically resurrected musicians parted company to embark on separate international tours, singing to fans old and new across the world who never imagined they'd be able to see these idols on stage again. Check your local listings magazine, because holograms don't tire easily and can tour indefinitely, as long as the money's still rolling in for their minders.

Drumming up ticket sales on its website, BASE Entertainment gushingly employs the language of perpetuity, tantalising prospective audience members with video clips of the Callas and Orbison holograms performing, corporeally indistinguishable from the members of the live orchestra backing them. 'Forty years ago, Maria Callas shrugged off the earthly binds of temporal life, leaving a bereft world longing for her presence, her performances, her voice,' says the blurb. 'Now she returns to the stage, in all her grandeur, confirming her immortality to stunned audiences.' Roy Orbison's pre-existing immortality has also been upgraded. 'His voice was always immortal,' sighs the blurb. 'Now the man himself is too.'[305]

When it comes to creating hyper-realistic holographic projections of dead stars and putting them on the road again, BASE Entertainment is far from being the only game in town. There are a lot of legends who've been taken suddenly and too soon, leaving their fans bereft, and more than one company is needed to cover them all. Hologram

USA has reanimated Whitney Houston and Billie Holiday, who were only in their forties when they died. Michael Jackson passed away in 2009 at the age of fifty, but when he appeared at the Billboard Music Awards in 2014 he didn't look a day over forty. So while death may be unfortunate indeed, especially when you go before your time, it's no longer an impediment to your effortlessly carrying off an active touring schedule.

There are other barriers, though, many of which will sound familiar from previous chapters. Technologically, constructing believable holograms is an uphill battle, but as audio and video digital editing and CGI techniques evolve, teams of specialised technicians are producing ever more cutting-edge versions of the old Peppers ghost projection trick, originally invented to feed the hunger of séance-crazed and Spiritualism-mad audiences of the mid-nineteenth century. The fact that they're not yet perfect, though, engenders instinctive psychological resistance. Three-dimensional, moving human figures that look incredibly realistic on one hand but that still have dashes of weird about them cause us to hit the brakes as we enter the uncanny valley. If press articles are any indication, this was particularly the case when dead stars first started appearing on stage. For example, back in 2014, some attendees at the Billboard Music Awards welcomed the spectral visit from the King of Pop, but a distinct segment of the audience was less sure, with many people expressing frank discomfort at seeing Jackson moonwalking across the stage once more. A journalist from the *Washington Post* described the illusion as 'digital formaldehyde' in an article headlined, 'Are holograms a creepy way to honor fallen icons like Michael Jackson?'[306]

In more recent years, consistent with the shifting location of the uncanny valley and the trend towards wondering if the dead continue to possess rights over their image and their privacy, objections are starting to take on an ethical rather than a psychological flavour. How might deceased icons have felt about being exploited post-mortem for fun and profit? How do we know whether a chanteuse like Billie Holiday would have liked the idea of digital reanimation, when she died in 1959 and could scarcely have imagined such a thing?

And what about all the legal complications? Who owns the rights to a famous person's name and image and likeness when that person is no longer alive? What about performance rights, composition rights, trademark rights?[307]

Where the latter questions are concerned, the answer is likely to be the estate of the deceased, the individuals inheriting the intellectual property rights. Far from placing obstacles in the way of post-mortem holographic reanimation, those who own copyright on fallen idols' work are queuing up to explore how a post-mortem tour may replenish their dwindling coffers. The CEO of Eyellusion, Jeff Pezzuti, has reported that his company is frequently approached by the estates of dead musicians, particularly when the album sales and other revenue streams start to dry up.[308]

If you're a cynic, a musical purist, or a post-mortem privacy advocate, inclined to decry this exploitation of dead musicians, well, Pezzuti thinks you're just being selfish. '[T]here's a lot of people that wanna see [heavy metal musician] Ronnie [Dio] that are younger or that never saw him,' he was reported as saying. 'Because you feel it's not right, or somebody feels it's not right, you're not allowing the younger audience or the next generation to be able to experience this music.'[309] Alex Orbison was reported to express something similar while standing outside the Rose Theatre, talking to a reporter from music magazine NME about how his mind had just been blown by seeing his late father onstage. If younger generations see his dad play 'live' and it inspires them to go out and make music themselves, Alex Orbison said, that's all that matters.[310] 'Some kid's gonna come and watch a hologram of my dad and go home and go, "That's what I'm going to do with life" . . . My dad went to a show when he was five years old and he left and said the same thing. That's the circle that we look for.'

For characters like Orbison and Callas, Houston and Holiday, Dio and Jackson, and for anyone who was rich and famous in their lifetimes, it isn't just their family and friends who care about their legacies and who yearn to hear their voices again. The uncanny reappearances of musical legends on stage are meaningful on multiple levels to multiple stakeholders, and Jeff Pezzuti and Alex Orbison aren't necessarily wrong when they theorise that these apparitions could spawn new generations of fans. Callas may have shrugged off the earthly bounds of temporal life, as the BASE website put it, but her hologram can afford to be even more casual about temporality, able to entertain future generations of opera lovers for many years to come. She may shape their tastes, too, and perhaps the digital twin of La Divina, abroad in opera-land, will threaten to eclipse living, breathing singers with new voices and fresh styles. That's one objection some opera lovers express, and it could indeed prove a real danger – Callas was, after all, exceptional.

But what about ordinary folks? What if you'd like to carry on having an impact in the world yourself, beyond just family and friends, even if you haven't been blessed with the stage presence of a superstar, or the voice of an angel, or the services of a really good PR agent? Granted, if you're a regular sort of mortal and not a legendary performer, perhaps it's still a tad early for you to be thinking in holographic terms. Discussing digital legacy with individual members of the public in 2018, at the Love After Death event in London, I was generally greeted with bemusement when I asked people how they'd feel about being reanimated after death, perhaps on specific anniversaries or special occasions, as a digital representation. Holograms aside, what if, despite being a relatively small fish in a huge digital pond, you'd like to have some kind of influence in the wider social sphere when you're physically gone? Thanks to the levelling power of the Internet, that can be arranged. In the online arena, you could be as vocal in death as you were in life but, to achieve that, you'll need to plan ahead.

Ω

Lucy Watts had always been a planner. By the age of eleven, she'd figured it all out – what GCSE examinations she'd be taking at fifteen, what A-level results she'd need to attend the medical school of her choice, even what she would do once she retired from a successful career as a doctor. It was a tender age to be setting and pursuing goals so fiercely, but she couldn't afford to relax. Lucy was harbouring a secret, so closely held that even her own mother didn't know how much she was struggling. She'd always had health problems, but by the time she was a pre-teen, her muscles were so weak and uncoordinated that she could barely lift her arms above her head or get up a flight of stairs. She couldn't pretend any more that it wasn't happening.

Flash forward to three years later, and Lucy was confined to a wheelchair, the root cause of her illness a frustrating mystery. She'd received a diagnosis of Complex Ehlers-Danlos Syndrome, a disorder that renders tissues fragile and joints hyper-mobile, but that was only a hypothesis that sort of matched her doctors' best guess. Nobody really knew for certain what was wrong with Lucy. They only knew that it was eventually going to kill her.

Unwilling to give up, Lucy was determined to wrest control of her fate back from her errant body. After all, she reckoned, a wheelchair didn't have to stop her from being a doctor. Ignoring the gloomy predictions that she might not live long enough to realise her ambitions, she kept her eyes firmly fixed on the path ahead. She cracked on with her studies, amazing her tutors and inspiring her peers in her quest to complete her GCSEs. While waiting for her results, she learned that she'd won a Diana Award.[311] Her tutors, bowled over by her dedication in the face of such difficulty, had made the nomination for the prize, which honours outstanding young people who create positive social change.

With a solid set of GCSEs in hand, she raised funds for a special wheelchair that would enable her to attend college, and she started studying the sciences, the next rung on the ladder to becoming a doctor. But Lucy's existence was rapidly becoming an interminable medical procedure. Bit by bit, the illness chipped away her hopes for the future. Unable to take food by mouth, she received nutrition directly into her small bowel. Sitting up made all of her blood drain down to her feet, forcing her to recline most of the time. At first, she was able to attend classes for an hour a day, but eventually even that became impossible. Her offline social life dwindled to her family members, one loyal friend, and the former riding teacher who brought horses round to stick their heads through Lucy's bedroom window. Her online connections, on the other hand, expanded to fill the gaps, including lacunae in her medical care.

As a less-visible patient, largely confined to her home, Lucy often found it difficult to get the attention she needed. Doctors in the National Health Service didn't generally make house calls and, in an overstretched system, it was hard for them to see past the patients who were physically sitting in the waiting room. But in early 2011, Lucy was losing weight rapidly and saw her consultant. Let's check you again in three weeks, he'd said, but then the follow-up was rescheduled, and three weeks were threatening to become three months. Anyone could see the weight dropping off – she was now less than 100 pounds – but less visible were all the internal organs that were failing, including her heart, her bone marrow and various elements of her digestive system.

Searching desperately for help, Lucy and her mother found a website for J's Hospice, a charity in their native Essex set up for young people who have life-limiting illnesses. J's sent out Bev, a seasoned hospice nurse, who took one look at Lucy and knew that she was going to die. She immediately jolted the consultants into action, and it quickly emerged that Lucy had stopped absorbing nutrients from her feeding tube. The tube was swapped for a nutrition line directly into her heart, and Lucy was pulled back from the brink. 'Actually having that information at your fingertips has a whole different meaning for me,' said Lucy. 'It's not just about my life socially. It's about my life in its entirety. I wouldn't be here without the Internet.'

Lucy might have been rescued from imminent death but, while her treatment changed, the condition was still life limiting, whether it would ultimately be the illness itself or an opportunistic infection that would kill her. Aware of these realities, Bev started talking to Lucy about advanced-care directives and end-of-life planning. It's a typical topic in hospice care, a discussion that usually focuses on what will happen close to the end, on the types of medications and interventions people do and don't want, the things that will make people feel most comfortable and most in control as they lie on their deathbeds. But Bev didn't play it that way. In an entirely matter-of-fact fashion, she asked her young patient what she planned to do with her life.

At that particular juncture, the question flummoxed Lucy. 'Everyone wants to inspire in some way,' she said. 'Everyone's got that in them, that they want to impact upon other people, benefit other people. We've all got something we want to do.' She'd always planned to do that through being a doctor, but she wasn't some perfectly healthy teenager. She had a terminal condition that had forced her to quit college. Her immune system was in tatters, her plans shattered. Psychologically, she was pouring her energies into accepting the collapse of her dreams, all that would never happen, all that she could never do.

Although Lucy's mother and Bev were exceptions to the rule, so many other people thought that she couldn't do anything at all, that such a life was barely worth living. Confronted with her illness, her complicated medical equipment, and the nurses and assistance dog that accompanied Lucy everywhere, the able-bodied people she encountered shuddered, and pitied. 'People see the medical, the physical. They see my tubes, my wheelchair. They see my prognosis. But they don't see me, necessarily, as a person. You get people who are like, *ugh*, I couldn't live my life like that,' Lucy said.

But then here came Bev, throwing a curveball, treating her like a person and not an illness and, furthermore, as though the future were her oyster. When she asked her young, dying patient what she wanted to do with her life, she unlocked something that Lucy herself didn't know was in her. 'It was like my subconscious screamed something,' said Lucy. 'It just came out. "I don't want to be forgotten. I don't want my life not to have meant anything." I was shocked that I'd said it, because I'd never consciously thought it.'

Her mother, hovering nearby, told her not to be silly, aghast at the idea that her family and friends could ever forget her. But that wasn't what Lucy meant. Of course she wanted to be remembered by those she loved, but it wasn't simply about that. She realised that after her body gave out, she wanted to be remembered for something *in particular*, not just by her family, but by people in the wider sphere. Lucy wasn't sure what that particular thing was, but a new determination had been unleashed. 'This seed was sown about being remembered. Not necessarily to be famous, or be anything special. [But] being remembered for *something*. Legacy.'

Furthermore, she didn't want anyone else telling her what that legacy would or should be. The word that Lucy used most when we spoke was 'control'. 'I've got so little control in my life, with my illness,' she said. We were sitting together in a room of her house, Lucy propped up in the reclining wheelchair that took up most of the room, her assistance dog lying nearby on the floor. I could hear one of her nurses bustling about in a room nearby. 'I can't control my death. But I can control how my life is looked back on.'

Lucy figured that if she left it to others to determine the sum of her life, there was too much danger that they'd underestimate all that it had been, all that it had meant. She'd experienced first-hand, too many times, people being unable or unwilling to believe that a worthwhile, wonderful, amazing life could be possible for someone like her. 'We are a judgemental society. We make snap judgements about people. My internal feeling is that my life is worth living, but my external feeling is that I want *other people* to know. Part of my legacy is, I want other people to see that actually, I had a great life. Despite everything. And because of everything.'

Only one person could deliver that message the way Lucy wanted: herself.

To control the message, though, she needed a medium that she could also control. In 2011, just after her most harrowing brush with death to date, she would have struggled to accomplish face-to-face conversations with everyone who needed to hear her message. In her acute state of recovery, attending events and doing speaking engagements wouldn't have been an easy proposition. But if Lucy didn't want to wait to have her voice heard, she didn't have to. For years, the Internet had brought the world to her room, and now she would use it to take her message out into that world, virtually at least.

At first, her blog was called Overcoming Obstacles, and her growing readership included medical professionals, young people, parents and families, including those with no experience with disability at all. She made friends well beyond the UK, in Iceland, Canada, America, Australia. After a while, as her mission became clearer, she thought of a better name for the blog, one that symbolised what impact she hoped to have in the world. 'My life has gained a lot of meaning through being poorly, not just meaningful for me, but for others,' she said. 'You can be that light to show people the way to go, no matter what they're struggling with.'

When I first met her, in late 2017, in a lift going up to a radio studio at the BBC, Lucy's Light had racked up over 265,000 views, and the more she'd been visible online, the more opportunities she'd had.[312] The accomplishments listed on her blog put the average CV to shame. A video for J's Hospice. Speeches in front of the House of Commons and at the Department of Health. Multiple media appearances for every major news outlet in Britain. Ambassadorships and other work for numerous charities. Authorship of guidelines that are read and put into practice by medical professionals across the nation. And then, in 2016, at the age of twenty-two, she was appointed an MBE – a member of the Most Excellent Order of the British Empire – for her services to young people with disabilities.

Maybe there's something about having a purpose that helps to stave off death. Having outlived her prognosis, Lucy had every intention of continuing to defeat the odds. As jam-packed as her schedule was, and with new projects and ideas percolating all the time, part of her didn't even think about her life as being short. She knew full well, though, that her fate was highly likely to cheat her out of a long life, to try and rob her of making her mark. If she wanted to have any kind of control over it, she had to think quite deliberately about what she wanted to leave behind, and what it would look like.

'It's funny, legacy,' Lucy said. 'It comes in different forms.' To her mother, she knew she would leave a legacy of love and memories. To her sister and any future generations, she wanted to pass down something more: knowledge, choice and control. Lucy was doing everything she could to find out if her condition were carried on her mitochondrial DNA, a genetic heritage her sister and any of her future offspring might share. 'My sister's children are very important to me,' she said. 'I might not live to see them. I just hope I've left them something that's not just my memory but actually something tangible.' She was in the midst of seeking more test results and a proper diagnosis, a legacy to nieces and nephews that she may never know.

While her mother and other family could never forget her, and while any further test results and diagnoses she receives would be concrete and transferable, Lucy realised that making a lasting mark on the wider world would take more work. The Internet was the answer to that, too, and with her usual forethought and determination, she embarked on a master class in digital estate planning. While this is the sort of thing for which the hale and hearty struggle to find space in their diaries, Lucy knew she didn't have the luxury of time. She identified Lucy's Light as the first order of business. Her existing posts, which she knew could inspire and help thousands more people after her death, would need to stay, and the domain would need to be cared for by someone. Given that her mother was less technologically savvy than she and, moreover, terrified and nervous about letting her daughter down, Lucy compiled detailed blog-maintenance instructions for her, complete with screenshots. She'd made sure that her instructions for all of her accounts included up-to-date passwords and had listed her wishes for each account. She'd already decided, for example, that the Twitter account she used primarily for work rather than socialising could be shut down without causing significant pain to anyone. Facebook, on the other hand, was a major conduit for her more personal connections, and she used it to maintain close friendships with people she'd never met face to face, people from around the world. Her profile would definitely need to stay and, she hoped, might even be used to live-stream her funeral.

Lucy foresaw a problem, though. She'd been told time and again that her immune system was incredibly weak and that she could die at any moment. When she was poorly, which was often, she was sometimes unable to communicate, and her friends had learned to discern when she wasn't well by observing her activity levels online. Lucy couldn't bear the thought of not having the final word on her life. What if, when the time came, she was too ill to get online and pull together her final thoughts, or even to dictate them to someone? Faced with the anxiety of that possibility, Lucy did what she always did, the same thing she'd done when people asked her how she managed to cope with such a restricted existence. 'I don't have a choice. Well, I *have* a choice,' she said. 'I can sit back and wallow in it or do something constructive.' Creating messages to be posted on social media after she'd died would give her back the sense that she'd left no stone unturned. Her life, her purpose and her legacy would be hers, and only hers, to define.

When the idea first occurred to Lucy, she'd never heard of anyone else doing such things. 'I sat there and I thought, why don't I write something that my mum can post after I've died, like a little eulogy, in my own words? And I thought, well, actually, what about a video? What about Facebook posts? . . . . I was quite assured in myself that that's what I was going to do. I didn't have a name for it, didn't know it was a thing, because I didn't know what I was doing *was* a digital legacy.' She got mixed reactions from family and friends, some of whom said it was wonderful, others of whom found it a bit weird. Then, through her various connections, Lucy happened to hear about a documentary on BBC3 called *Rest in Pixels*, and suddenly saw that she was far from alone in her thoughts.[313] There was a world of people out there who didn't think that the concept was strange at all, and 'digital legacy' entered her vocabulary for the first time.

Some of the folks also thinking about it were developers like Steve and Peter, from Chapter Six. Although they weren't marketing My Farewell Note and GoneNotGone specifically to terminally ill people, their services had certainly found appeal amongst those like Lucy, people who knew that death was close. She could have used a service like that, but these seemed only designed to say goodbye to specific people with whom the deceased had had a close relationship, or to help the deceased continue being present in those individuals' lives.

Lucy, who wanted to leave no stone unturned, knew there were far too many stones for that kind of messaging service to handle.

By her early twenties, thanks to the Internet, Lucy had become a public figure, a friend to hundreds and an inspiration to thousands, and she wanted her final messages to go out to them in her own words and on her own terms. She wanted to sum up her life for *everyone*, not just for a fleeting moment, but for far longer. For all of the people who knew her, she wanted to remind them how important they'd been, how much she'd learned from them, how much she loved them. For all the non-disabled sceptics out there, and all the people with disabilities who also suffer but whom she'd never met, she wanted to warn them never to underestimate the worth and potential of a life like hers. Right here, she wanted to say, is the proof of all that's possible for someone like me. What could help her do all that?

Luckily, the *Rest in Pixels* documentary heavily featured James Norris, the software developer who'd launched the non-profit organisation Dead Social at 2012's SXSW, his contribution to the 'death tech space'. In an unusual twist on the typical posthumous-messaging offering, James had designed Dead Social to deliver messages through the same social networks that Lucy used, every day, to connect with the wide world outside her bedroom.

Ω

Having lost his dad at an early age, James Norris thought a lot about death when he was growing up. Listening to all of the talk about Heaven and Hell at his Catholic school, he mused about what kind of afterlife he and his loved ones might end up in. He felt a bit guilty about the skulls on his Guns N' Roses T-shirts, wondering if he should be wearing that kind of imagery when his father had just died, but he found their music comforting.

Skateboarding and listening to rock music was what normal teenagers did, but at the same time he wasn't sure if his friends spent quite as much time as he did pondering the songs that they wanted for their own funerals. It wasn't that he was suicidal, he was just planning ahead, and he knew that he definitely wanted Guns N'

Roses to be on heavy rotation during his final journey on earth. Unlike the teenaged Lucy Watts, the adolescent James Norris didn't have a terminal illness, nor a particular message he wanted the wider world to hear. To him, it was mostly important that his funeral playlist felt right – it was the send-off that preoccupied him, not the legacy.

James isn't entirely sure why it was so psychologically crucial for him to feel in control of his own funeral songs. Perhaps it was because he had found so much solace in the music of Guns N' Roses when his dad died. In any case, because it was important to him, he thought it was important for others to consider it too. I spoke to James before I talked to the superintendent of the City of London Cemetery, but I rather suspect that I know what he'd think about one of the tales that Gary told me, the one about the crematorium manager's censoring a deceased young person's favourite song to protect the serenity of the cemetery and the potentially more delicate sensibilities of some of the mourners.

James feels sufficiently passionate about final songs to include questions about them in the periodic questionnaires he puts out for the Digital Legacy Association, even though one could argue that the notes wafting through the air as the mourners assemble at the funeral home or crematorium don't have that much to do with one's digital footprint. 'Have you thought about which song(s) you'd like to be played at your funeral?' asks one of the items on the 2017 survey, wedged incongruously between questions about plans for your social media profile and whether anyone knows the password for your mobile phone. 'If you have thought about which song(s) you'd like to be played at your funeral, have you told anyone?'[314] In his early twenties, though, James stumbled across something that inspired him to broaden his thinking about legacy. He was working as a social media manager for a digital marketing agency, and in the course of doing research for some charity projects, he discovered an intriguing video on YouTube.

The advert opens with a startled flock of birds – maybe ravens, maybe crows – taking flight, silhouetted black against the grey background and the steeple of an old church. They've been disturbed by the comedian and TV presenter Bob Monkhouse, who's strolling among the tombstones and looking remarkably well for someone who's been dead for some time. The soundtrack to this graveyard scene is tinkling and jolly, and Monkhouse's ghost launches immediately into his familiar comic schtick. 'Just when you thought it was safe to turn on your TV again, here I am!' he quips. He pauses by his own grave marker – Bob Monkhouse, it reads, died 29 December 2003. Monkhouse's ghost gives some surprising statistics about how many men were killed by prostate cancer in Britain – one every hour, 'that's even more than my wife's cooking! Hahaha!' A screen pops up to say that prostate cancer cost Bob his life and admonishing the viewer not to let the same thing happen to them. 'I'd have paid good money to stay out of here,' sighs Monkhouse, shown sauntering resignedly away from his own grave and disappearing into thin air. 'What's it worth to you?'[315]

While James had never seen this advert before, it had originally aired on British television in 2007 as part of a prostate-research fundraising campaign during Male Cancer Awareness Week. Whether it would have made the same impression on someone who hadn't seen their own father sicken and die, it's hard to say, and if James hadn't been a social media guy at a time when social media was a burgeoning phenomenon, he might not have thought much more about it. As it was, though, a massive lightbulb clicked on above his head. Look what this man has done, James thought. Anticipating his death, using the power of his celebrity and the mass medium of television, this famous late comedian had delivered a highly personal, humorous-yet-serious sucker punch from the great beyond, in his own inimitable style, as though he were truly speaking in real time. 'His one-liner jokes in there were very, very true to him as a comedian and to him as a person,' James said. 'It was quite clearly a Bob Monkhouse sketch. It was poignant, because it appears that he created it in the context of it's only being shown [four] years on, after his death.' Monkhouse was a celebrity, and when he died in 2003 you needed to have celebrity or at least connections to be on TV, but it wasn't 2003 any more. Everybody could do the same thing as

Monkhouse had done, James thought, using social media not just to say goodbye to loved ones, but to get meaningful and important final messages out to *everyone*. It was the catalyst for Dead Social.

The idea of creating personally meaningful messages for posthumous delivery held massive appeal for James, a seemingly natural extension of his long-time fascination with choosing the last song. Partnering with another developer, he created a system that, with the help of a trusted digital executor, allows people to post pre-written messages on Facebook and Twitter, according to a schedule set by the deceased. Wanting it to remain completely free, he designed it as a non-profit and capped the number of subscribers at 10,000, in principle, which he'd since soft-heartedly allowed to creep up to 13,000 in practice. Such was James's captivation with his concept that he had no doubt that everyone else would find the concept alluring as well. 'I thought, after launching it, everybody's going to want to do this, because it's something that I want to do,' James said. But, like so many other developers in the 'death tech space' have also discovered, to their cost, particularly those who had fully expected to make their fortunes, James quickly twigged that this simply wasn't the case. Just as not everyone shared his enthusiasm for continually refining one's funeral playlist, not everyone had a deliver-after-death message burning a hole in their pockets.

It turns out that the person who inspired James's lightbulb moment, Bob Monkhouse, didn't have one either. Just as Michael Jackson embarked on his 1997 HIStory tour with no idea that video footage of his performances would be mined years later to produce a holographic resurrection for the Billboard Music Awards, the late comedian had no cunning plan to deliver a health-related public service announcement to the men of Britain from beyond the grave. A combination of archival footage, a body double, a sound-alike voiceover artist, and computer magic created the illusion of Monkhouse giving his prostate-cancer message. As his widow said to the BBC, it's probably the case that he would have loved the advert, but he certainly didn't plan it.[316]

Forced to acknowledge that Dead Social didn't have the broad appeal that he thought it would – at least, not yet – James did something similar to Marius. He rethought things, and he diversified. He formed the Digital Legacy Association, an organisation providing training, workshops, advice and consultation about digital assets and digital legacy, and that would prop up the non-profit Dead Social along the way. As part of the DLA's work, James hounds companies like Amazon and Apple and Facebook about quality and transparency in their processes, works to convince sceptical and laggardly policy-makers and regulators about the importance of considering digital assets and legacies, and provides training and resources to eager palliative care workers who know full well, from being on the ground, how important online connection and digital legacies can be to people like Lucy.

Eventually, things got busy enough for him to need a team of people to help him, including Lucy as Young Person's Lead, Vered as Middle East Lead, and even myself as Psychology Lead.[317] He hasn't forgotten about Dead Social, though, and in 2018 he was preparing to launch an updated platform, My Wishes.[318] James just doesn't feel that he'd be doing everything he could to help people plan for their digital afterlives if he weren't also helping them use their social media channels, so much a part of some people's lives, to have the last word.

Lucy's collaboration with the DLA has made her consider her digital legacy even more thoroughly, and it would be a challenge to find anyone who has put their digital house in order more tidily than she. It might surprise you to learn, then, that even this ultimate planner hasn't filmed her intended final video messages yet, and neither has she written the posthumous blog posts that she wants her mother to release at intervals after she's gone. Lucy may know precisely what she wants her lasting mark on the world to be, but she always runs up against the same problem that I encountered myself when I was struggling with the concept of final messages for My Farewell Note and GoneNotGone. Even as her mystery disease progresses, even as she races against the clock to find her true diagnosis while the nutrition that keeps her alive drips through the line into her heart, she's not ready to record a final message yet. She's immersed in her life. 'I've half-written a lot of blogs. But I keep changing my ideas,' Lucy said. 'I do struggle with that ... I'm not there yet. I shouldn't be here now, and I could die at any point. Do I do it today, just in case it happens? Or do I wait until next week? . . . . but I'd be writing the end while the middle's still happening.'

Even if Lucy never gets the chance to write the ending, if she misses that opportunity because until the very last moment she feels like the middle is still happening, she can be comforted with this. Gathered together, neatly organised, it's all there, the digital autobiography that she's been compiling all the way along: her published works, the transcripts of her speeches, her blog posts, a list of her accomplishments, the story of her life, even the link to a blog kept on Facebook 'by' her beloved assistance dog, Molly. She knows that the final messages that she hopes to leave, what she calls 'the update at the end of the book', aren't her legacy. Her legacy is everything she's done. She's been her own online biographer and archivist for the past seven years and counting – constantly tending the light that she wants to be for others in the world, until, one day, the time will come to hand the care of that flame over to someone else.

Ω

If Jason Noble ever had a mission like Lucy's, to be some kind of guiding light for others in the world, he's highly unlikely to have ever put it in those terms, and even more unlikely to have said it on the Internet. Compared with Lucy, Jason's digital footprint is spartan and fragmented, unsurprising perhaps for someone who grew up listening to LPs, making cassette tapes, and drawing comics and 'zines with ink on paper. The music from his various bands can be downloaded from iTunes, and the pieces that he wrote for the Louisville arts and culture publication *LEO Weekly* appear online, but his social media traces are sparse indeed.

His friend Greg King described Jason's scant online presence as weird and funny, adjectives that capture well-loved elements of his friend's offline persona as well. Because Jason didn't really do social media, Greg was completely unaware that he was on LinkedIn – until, that is, the day the site suggested that becoming acquainted with Jason Noble might prove useful for his long-time friend and band-mate. 'Greg, staying in touch with valuable contacts can help you in your career,' it said. 'Quickly connect to some people we think you know.' LinkedIn didn't know the half of it, and Greg was only too keen to oblige, although career advancement was the last thing on his mind. Clicking on the yellow box to 'Connect with Jason', he wondered what he'd find.

Summary: Humanoid – 1 percent difference in chromosomal architecture from most non-human primates. Fan of 'The X-files' and anything with dusty corridors or flashing lights. Specialties: musical sounds/Comic Drawings/Nocturnal Ramblings/Night Terrors/ Film & Audio Production/Reptile Body Temp.

Experience: Musician, Shipping News/Young Scamels/ear X-tacy. Currently on medical leave but still working (slowly).

Overview: General operations . . . garlic and crypt supervision. Sometimes uses stakes/silver bullets. Use of ancient texts/white oak shavings/blunt tools. Willing to work nights.[319]

Greg was stunned. Jason had been dead for months, carried off at just forty years of age by a bacterial infection, his immune system weakened by multiple rounds of treatment for a rare form of cancer. Greg was also in hysterics. 'This weird digital algorithm brought this little moment of joy into my suffering,' he said. 'It allowed me to connect with him again and see something of him that I hadn't seen, like hearing a joke from him after not having heard from him in months. And it was like a very private moment . . . I really kind of loved it.'

This moment was particularly meaningful because, no matter how uncanny this algorithmic sneak attack was, it also provoked an entirely familiar sensation, one that Greg had feared might stop with Jason's death. In the memorial tribute that Greg wrote for Jason in August of 2012, he described an occasion on which Jason had been visiting him in New York City. At the close of the visit, while Greg was getting ready for his day and Jason was making preparations to leave, Jason sketched a picture of Spider-Man and secreted it among the flat files in Greg's room. 'Just a bit of surprise Jasonalia for me to find someday,' Greg wrote. 'Someday' didn't prove to be too far off. Mere moments after Greg had stood with his friend on the stoop of his Brooklyn duplex, giving him a hug goodbye, a mysterious force drew his attention straight to the file in question. 'I sat at my desk, just ruminating on the nice visit I'd just had with him, and my eyes fell on the flat files. Something inside me compelled me to open a drawer, and literally the first one I opened had the drawing in it,' Greg told me. In his memorial tribute to Jason, Greg said that he 'lived for those moments, the strange brainwave connection we could have, just the honest and often goofy conversation we kept airborne between us since that first day in high school, a collaboration of living filled with uncanny coincidence, timing . . . . and superheroes.'[320]

Greg might have taken himself out of Kentucky, opting for larger urban centres more conducive to a career as a filmmaker and artist, but it would take more than a move to New York or LA to take Kentucky out of him. Louisville – the place where he'd grown up and attended high school with Jason – is part of the fibre of his being. When Jason died, Greg – reeling from a 'terrible emotional bomb' – was living on the West Coast, isolated from the community he most craved at that moment. 'My close friends in LA, they don't really have an understanding or connection to my Louisville community, and my Louisville community is incredibly important to me,' he said. 'It's so part of my identity, it's hard to fully describe to people.' Needing to make contact with anyone from home, any folks who'd also known Jason, he searched them out, spending the next day with people he hadn't seen in years.

But Louisvillians like them weren't the only ones in mourning. The various bands Jason had spearheaded over a fifteen-year period – Rodan, Rachel's, the Shipping News – had fans around the world. In Jason's obituary, *Rolling Stone* described Rodan as having had a profound influence on the post-hardcore music of the 1990s.[321] Of the hundreds of people who followed the progress of Jason's illness and treatment on the medical-updates website CaringBridge, some of them were music-lovers from places like Italy, posting their messages of support in foreign languages.[322] I know all this because, concerned about his illness and saddened and moved by his eventual death, I had been closely following Jason's CaringBridge page too, and I felt rather uneasy about that. Was I being like the central characters in Hal Ashby's cult 1970s film *Harold and Maude*, with their shared hobby of showing up to strangers' funerals?[323] Even though it was on a public forum, was I doing something unseemly just by being there, much less contemplating whether I should post a comment too?

I reassured myself, though, that it wasn't like I didn't *know* Jason. I was from the same place. I had been acquainted with him as a teenager. I had dated one of his friends, ridden with him on a few car journeys, visited his house. He appears in some of the photographs stored in one of the many shoeboxes in my attic. In later years, on trips back to America, I had chatted with him now and again at ear X-tacy, the mildly famous and now defunct alternative record store where he worked, but although I wish we had been better friends, for the most part – just like the fans who had never met him – I experienced his talent and spirit from a remove, through his music.

Given the slightness of our association, I wondered if I were being a voyeur, a dark tourist. After so many years without contact, did I really know him well enough to warrant an emotional stake, an experience of genuine concern and sadness? On the other hand, I felt unapologetic for caring about someone who had always treated me and everyone else with kindness and generosity. He was another human – a fellow 'biped', as he put it on his LinkedIn profile – who had been an irrepressible font of creativity in health, and a lion of courage in suffering. The kind of discomfiture that I was experiencing isn't unusual in the age of social networking. Nearly everyone understands and sanctions publicly expressed grief when it's clear that the grievers were close to the deceased, but when less intimately connected people voice their sorrow, it gets trickier. When the object of publicly expressed sadness is a celebrity or some kind of cultural icon, someone who might have had a huge impact on your life but who wouldn't have known you from Adam if you met them in the street, the 'grief police' are fast on the scene. You didn't know him, they might say, stop this inappropriate emotional rubbernecking,[324] move along! You weren't a proper fan, they might say. Stop disrespecting the memory of my idol and leave the mourning to the *true* fans, like me!

'When a celebrity passes away, their fans experience the loss of what psychologists call a *parasocial* relationship,' explains one group of researchers, including Jed Brubaker, who analysed grief-policing comments on Facebook after the 2016 deaths of Alan Rickman, Prince and David Bowie.[325] Parasocial relationships are described in their paper as 'a one-sided, mediated relationship that many people may experience with similar emotional strength to personal relationships'. The online environment may have contributed significantly to the development and strength of those parasocial feelings, and online communities may be the absolutely perfect opportunity for the community of fans to easily come together in their grief, but without well-established norms around when, where, why and how you're 'allowed' to grieve in such a situation, it's always possible that someone will get upset with how you're expressing yourself. Jason Noble, I have no doubt, would have either recoiled from the idea of himself as a 'celebrity', or found it hilarious, but nevertheless my uncertainty was definitely a variant of the modern-day 'fan's dilemma' in the face of death. Was it my place to feel, much less express, my sadness?

But yes, it's OK, I decided. You did know him first-hand. You can just about scrape into the category of people who will, as one memorial tribute's headline read, 'Carry Him Forward'.[326] In some small way, at least. I was unable to resolve my unease quickly enough to post anything on CaringBridge, eventually opting for something more passive, lurking in a place where his community of mourners and fans continue to carry him forward, a memorial Facebook group.[327] When Greg first heard that the Remembering Jason Noble group had been set up, he fell upon it hungrily. 'I just lunged for it. It was like a buoy or anchor or something in a storm,' he said. 'It was addictive, just a gut response of needing to hear stories and see things and connect with people.' Living such a long way from Louisville, it felt especially vital for him, and it still does. To this day, on the memorial group, Greg repeatedly finds something he endlessly craves. When someone from that community posts a memory, or a photograph, he experiences little shocks of wonder and love, similar to what he felt when he found the Spider-Man drawing in his file, or when LinkedIn connected him with Jason's tongue-in-cheek professional profile.

'There are dozens of things that I have no idea about personally, because why would I?' Greg said. 'He gave a friend a drawing, a co-worker at work or whatever, and that person will post a drawing or some sort of memory, and I love that so much, I just eat it up. I kind of can't have enough of it, because it's just helped me reconnect with him and it feels kind of new. It puts a little bit of story where I didn't know something existed. So it's like continuing his life.'

Despite all of the joy and comfort that Greg has received from the Facebook group, though, he's been feeling uncomfortable too. Unlike me, his unease isn't about his right to be there, or to express his feelings. He was one of Jason's best friends, and his status as a griever was beyond question. I hadn't spoken to Greg in years either, but shortly after all of the publicity around Cambridge Analytica's use of Facebook members' data, I spotted a comment from him on Jason's memorial page. It stood out amongst the usual posts in that it wasn't a photograph, a video, a link to something Jason would have appreciated, or one of the messages addressed directly to Jason that many of his friends often made. Instead, the post expressed a dilemma that Greg was wrestling with, and it was directed to the other members of the community.

'Friends,' wrote Greg in the post, 'one of the main reasons I keep up my Facebook account is access to this memorial page, but I'm thinking of moving on from Facebook. However, I want to archive this page, and have started in my own painstaking, screen-grabbing way.' He'd been doing the same thing with the emails from Jason, still in a Yahoo! account that Greg didn't want to be wedded to for ever either. He was busy with work and family, and life kept getting in the way of all his laborious efforts to capture the existing content of the page one screenshot at a time. Did anyone on the page, he wondered, know of a better, more time-efficient way to get this done? In the flurry of answers that followed, there was a range of responses – practical suggestions about download services, creative ideas about producing a physical book or constructing an alternative archival website of Jason's work and life, and blind panic from people reading the post too quickly and mistaking Greg for an administrator who was intent on closing the group down.

Greg was quick to reassure them – he didn't have the power or desire to do that; he was just someone who was considering leaving Facebook. He was tired of the vitriol, the one-sided conversations, the depressing politics, the lack of true discourse. At the same time, Facebook connected him to his Louisville community and allowed his mother to easily keep tabs on her little grandson, and, despite his gripes, he was often grateful for the outlet and the format afforded by the platform: the combination of visual, audio and text material; the connection with far-flung communities. Greg's relationship with Facebook is the same love/hate one that many of us feel, and he felt torn, but he was sure of one thing: if he did decide to pull the plug, he wasn't going to do it without first finding a satisfactory way of archiving the content on Jason's memorial page. His quandary was a perfect example of the hypothesis I'd been reluctant to propose to Jed Brubaker, the idea that Facebook might in some small way benefit from retaining living users who were reluctant to lock themselves out of the cemetery.

I recognised what Greg was expressing, the modern tyranny of being 'always on', the Faustian bargain of social media, and the yearning for a simpler, more analogue time. But the possibility of his leaving Facebook wasn't the only thing feeding Greg's impulse to physically archive online material that was connected to Jason. 'I'm old school,' he said, at the conclusion of his post on the Facebook group. 'I want a physical copy. I want my collated, three-ring binder full of print-outs of this here JBN page.' Referring to the kind of sci-fi machine-against-machine showdown scenario that Jason surely would have appreciated, Greg said that he fully expected the computers to be killed off by the robots within his lifetime. When we spoke voice to voice, he reiterated that fear, but in a more serious vein. Working in digital media as a filmmaker and editor, he has considerable insider knowledge of the fragility of digital technologies, and that knowledge fuels his insecurities about any digitally stored material. Greg knows full well that any equating of digitally stored material and immortality is completely illusory.

'I'm haunted by this problem ... we've gone tapeless now. Everything is filed on [memory] cards or drives,' he said. 'If you have money and your private studio, then there's mainframe backups and stuff like that, but if you're just trucking along and making your stuff and shooting photos, you're relying on spinning discs, and now more solid-state stuff, but you're relying on things to preserve that media in a way that's not built to be long-lasting.' And even a well-established website like Facebook, he said, just didn't feel *concrete*. What if something happened? What if the administrator closed it down? What about when Facebook, and websites as we know them, cease to be a thing?

I remember Marc Saner of the World Wide Cemetery talking about our baby Internet, and Gary at the East London Cemetery saying that, despite Victorian fantasies of perpetuity, stone memorials have always been designed to fall down. I remember Michaela, whose late boyfriend had been an artist and filmmaker like Greg, and who had lost Luka's entire body of work when his next of kin took away the laptop that later broke. Despite Luka's profession and his awareness of how he ought to safeguard his work, its contents weren't backed up. He was one of those people just like Greg describes, like so many of us, just trucking along and making our stuff and shooting photos. More than once, I've heard John Troyer of the Centre for Death and Society admonishing his audiences not to place so much trust in their digital storage devices, with their frailty and potential for technological obsolescence. If you have photos that are important to you, he says, 'print 'em out. Right now.'

Of course, we know that physical memories aren't always failsafe either. A small part of what made it so painful for Rachel when she lost access to and control over her daughter's Facebook account was that, years earlier, her car had been burgled of the entirety of its contents during a house move. Those contents included precious physical photo albums containing mementoes of her late daughter's childhood. To assuage his own insecurities over continued access to Jason's memory, Greg felt compelled to chew through reams of paper printing out emails and Facebook posts – a mirror image of my own stress about whether I should go through the laborious process of making digital copies of the five volumes of my grandparents' World War II letters, currently arranged in three-ring binders – the same storage solution in which Greg was seeking a sense of security.

But Greg wanted to do more than just preserve material associated with Jason for his own use. He wanted to help ensure Jason's memory would be carried forward for others too. An idea proposed by someone on the memorial page seized his attention, a suggestion so simple and so beautifully appropriate that Greg was rather amazed that he'd never thought of it. 'I was like, holy shit, that's perfect. That's perfect. That's so Jason. He would *love* a book. Part of me is like, that is going to happen . . . . I'm going to make that happen. Maybe not this year or next year, but . . . that is an idea that I can latch on to that just seems so right to me.'

So right, and Greg is so well positioned to do it, but there's yet more reason for him to hesitate. Amongst the things he's learned in the past few years are that time is precious and can be short, and that fate can rob you of the life and legacy you could have had. Jason's death isn't the only thing that has taught him that. In another spooky coincidence, of which he and Jason had experienced so many, a year after Jason was diagnosed, Greg's fiancée was diagnosed with the same cancer. It was a development that could have ruined any chances for them to have children had they not moved swiftly to undertake egg harvesting before the gruelling toxicity of chemotherapy began. A month after their son was born, Greg himself was diagnosed with cancer, although for once it was a less rare type. Although he's recovered, Greg's not about to take his life for granted, and in fact that's what makes him second-guess his impulse to archive and perpetuate Jason's legacy in the world.

'By my nature, I would love to take a year or two and just archive Jason's life and work,' Greg said. '[But] I've had some real difficulties personally in the last few years in my identity as an artist, and I can't talk to the person I talk to the most about that. The way to deal with that, in a big way, is to keep making work, keep making new work. I would know exactly how to do an archive, to collect his work, but . . . would that be helping in trying to advance and continue my own artistic practice, which is something that he was a huge supporter of and benefactor of, a supporter of? He believed in me, you know what I mean?'

It may not exist online any more, but compulsive archivist Greg is able to send me screenshots of the profile that LinkedIn connected him to sometime after Jason's death, the moment that felt like such a personal communication, a cheeky posthumously delivered in-joke between him and his long-time friend. It's clear that Jason either created or updated the profile after he had already become ill, given references to 2011 creative projects and his being on medical leave. At the very bottom of the page, under a category heading of 'Additional Info', is a curious little snippet, blue and underlined on the screenshot as though it might have been a hyperlink.

Interests: The 'unquiet' grave.

When I emailed him about this, Greg replied to say that he didn't know what it might have meant. He thought that Jason probably included it as just a bit of 'spooky humor'– he was always employing the language and imagery of horror, he said.[328] Given my lightly social but mostly parasocial relationship with the late musician/artist/polymath, I'm inclined to defer to the judgement of one of his closest friends.

'The Unquiet Grave', though, happens to be the title of a well-known old English song. Far from being so archaic as to be only known to musicologists, it's still beloved today by folk singers.[329] While it has many variants, the gist is always the same. A young woman lies dead in the ground, but the young man mourning her weeps unceasingly, never leaving her grave. Unwilling to accept that he will never see her again, he neglects to take advantage of his own life. After a year and a day, the dead woman decides that it's time to speak up. Her voice issuing from beneath the earth, she tells him that in the garden where they used to walk together, even the finest flowers are all eventually fated to wither and die. If you were to join me now, she said, our hearts would simply rot away together, and that is rather a waste of whatever time you have left. You'll die one day, soon enough, but in the meantime, you'd best live your life. 'So make yourself content, my dear/'Til God calls you away.'

As driven as he is to remember Jason in both art and life, Greg still hopes to find a project that combines new work and Jason's legacy in one. Perhaps he'll make a film, he thinks, inspired by Jason, but as much about his own art as his friend's, as much new as old. Just like Jason would have wanted.

Ω

Especially given that Greg himself had experienced cancer, I felt moved to ask whether he consciously thought about his own body of work as 'legacy stuff'. He didn't miss a beat, as though it were obvious. Of course he's influenced by his own experiences of illness and loss, but it's more than that. *As an artist*, Greg said, he automatically thinks of things in legacy terms. Just as his now-and-future children are a part of his legacy, so are his creative works, and his hope is that one will support the other, that his children will appreciate what he has made in his life and will ensure that his paintings and drawings and films will have life when he is no longer here.

Works of art are born out of the minds of artists as surely as children are born of their bodies, and although art and artist may exist independently, they are inextricably linked. Both will only remain vital through communication, through communion. To live, the playwright's play needs an audience, the author's book a reader, the artist's painting a viewer, the musician's song a hearer. For him and Jason to live on, Greg can only hope that the devices on which their works are stored won't fail, and that Greg's children won't keep his art to themselves but will continue sharing it with others. As long as the creations live in the analogue or digital world, the creators live on too.

When the video for David Bowie's 'Lazarus' was released, its full implications were only made clear a few days later, in the 10 January 2016 official Facebook post that announced his death.[330] The video was indisputably Bowie's final communication to the world – unlike the Bob Monkhouse public service announcement, Bowie knew exactly what he was doing. Only he would deliver his message to the world – *I've got drama can't be stolen*,[331] he sings – and its content and timing were certainly true to how the public Bowie had always lived. It was a fitting conclusion to an extraordinarily curated and stage-managed existence. In the video, a hand reaches up from underneath his bed to pluck at his hospital gown, his face is like a death's head, the imagery and lyrics are rife with both explicit and oblique references to impending death. But suddenly Bowie is out of the hospital bed, sitting at his writing desk, gaunt but fizzing with energy. He rips the cover from a fountain pen and glares fiercely into the middle distance, his eyes firing with ideas, alight with inspiration. A lightbulb moment hits, and he jolts as though electrified, jabbing his index finger into the air, eureka! How can this be the end, when the middle is still happening? He hunches over, scribbles frantically, pauses to think again, returns to his labours, until, finally, it is over. Bowie collapses over the table, the ink trailing off the page, over the edge.

So little time, so many ideas, so much art, so many fans with their eager ears. So many watching, on their laptops and mobiles, in disbelief and with tear-streaked faces, as their idol recedes into the darkness of a Victorian wardrobe and pulls the door shut, signalling his return to some Narnia-like land where, surely, a creature such as Bowie must have come from in the first place. Fans everywhere felt that they'd received a communication directly from his lips, reassurance and solace in their grief.

Look up here, Bowie sings, in his unmistakable voice. *I'm in heaven.*[332]

Chapter Eight
Remembering Zoe

On 10 January 2016, a little girl called Zoe was holding a birthday party in David Bowie's native England. Zoe shares a middle name and a set of mitochondrial DNA with her great-grandmother on her maternal grandmother's side, Elizabeth; from both of her parents, she's inherited a keen appreciation of music. Her mother and father's tastes were shaped by their own parents' LP collections, and whatever was on the radio while they were growing up in the 1970s and 1980s. Their old cassette tapes, still around and slowly decaying in shoeboxes in the heat of the attic, are full of songs with the first few seconds chopped off – it took time to press 'record' on a tape machine, back in the day, when you heard your favourite track hit the airwaves. If you missed the DJ saying the name of the band, it could take days to figure out who was responsible for the song, so that you could head down to the record store with your pocket money.

Born into a digital era, with a Sonos system and a Spotify account and access to YouTube at her fingertips, Zoe had experienced no such impediments on her own journey of musical discovery. When she decided that she wanted a Bowie-themed birthday party, her proud parents were keen to oblige. On the big day, a DJ ran through the birthday girl's playlist of chosen songs, ranging from the 1960s to the present day. As Bowie lay dying at his home in the Catskills, far away across the ocean, dozens of children and their parents danced in a church hall in London, wearing band T-shirts, with glitter tattoos on their arms, and rock-and-roll paint on their faces.

ADMIT ONE
SEC:VIP ROW:10 SEAT:01

ST. JOHN THE BAPTIST CHURCH HALL
CHURCH LANE/HIGH ROAD

SUNDAY 10TH JANUARY 2016

1 - 4 P.M.

RSVP TO ELAINE BY 5TH JAN.

FOOD, SOCIALISING AND FACE PAINT
FROM 1 P.M

DISCO
FROM 2 - 4 P.M.

LET'S ROCK!

ZOE IS TURNING 6!

The next day, my daughter was shocked to see Bowie's photo spread across the front of the newspaper, a full-page image of the artist from his Aladdin Sane era. The jagged red-and-teal lightning bolt slicing across his face was the same as the one that, only the day before, had appeared on Zoe's own face, on her T-shirt, and even on the icing of her birthday cake. As she read the headline out loud, more than once, she kept skipping over the word 'dies'. *David Bowie . . . of cancer, age 69.* 'Maybe,' she said slowly, 'his family will carry on the music.' As though trying to get to better grips with the concept of ageing and death, she requested particular videos on YouTube. 'Now could we see one from when he was young?' she asked. 'Now could we see one from when he was old? Now could we see one where he's young?'

Logging on to Facebook again later that day (where, as is so often the case, I had learned about his death to begin with), I saw that all the comments beneath were about how Zoe had been the first person who sprang to mind when people heard. Such was the extent to which this particular network of friends, my friends, had come to associate my daughter with Bowie. 'When I heard the terrible news, she was the first one I thought of and I was wrecked,' said one friend. 'David Bowie's only regret in life will have been that he didn't meet Zoe,' said another. 'Hopefully inspiring another generation of kooks and loons,' said another.

'Have a look at "Lazarus",' said someone else, an acquaintance who had never met my daughter but who had also commented that this was likely to be her first major experience of loss. 'He left her a goodbye message.'

Ω

Before the digital age, nobody thought too much about the informational privacy of the dead, but all of the persistent, identifiable, hyper-personal data sticking around online is certainly making us think about how digital legacies should be managed, even if lawmakers and regulators remain unconvinced about fully embracing the notion of post-mortem privacy. But the dead are far from being the main entities whose privacy rights once received short shrift but are now being rethought. About another group, regulators are feeling far less ambivalent.

Who else – besides the dead and the otherwise incapacitated – haven't traditionally had much choice or power over how they are represented to others? Children. The new rights to privacy being granted them by law are coming into being because of the digital age; more specifically, because of their parents' use of social networking sites. The European Union's latest data protection regulations don't have anything to say about the data of the dead, other than suggesting that the member states deal with that conundrum, but the 2018 GDPR has a lot to say about kids. Minors, it states unequivocally, are data subjects too, who have the very same rights to control their information as adults do.[333]

Of course, the problem is that by the time a child gains the verbal skills, awareness, insight and understanding to grant or withdraw their consent, they will likely already have a substantial digital presence. In 2017, 92 per cent of two year olds in the United States already had digital footprints, in many cases dating back to before their births, for sonogram pictures are shared online in a quarter of all pregnancies.[334]

Doting parents everywhere have long loved chattering to others about their darling offspring, showing off their photographs, and waxing lyrical or getting competitive about little William's particular cuteness or little Sophia's superior accomplishments. The reach and extent of information disseminated via social networking, however, is another level. You may have come across a term that describes the online behaviour of many a fond mother or father: *sharenting*.

As one legal skills professor in the United States puts it, 'When parents share information about their children online, they do so without their children's consent. These parents act as both gatekeepers of their children's personal information and as narrators of their children's personal stories.'[335] Anyone who discloses personally identifiable information about their child to a mass audience is sharenting, although the practice varies considerably in nature, frequency and platform. Some parents write blogs and use Instagram to document and even monetise the joys, laughs, trials and tribulations of pregnancy and parenting; for some, often mothers, their 'mummy blogs' are sprinkled liberally with product promotions and are not just hobbies, but primary sources of income. On the opposite end of the sharenting spectrum, some parents may merely post the occasional photograph on a Facebook profile, with a limited audience of close family and friends, or post photos of the little ones privately on WhatsApp or a shared iPhoto album, controlling the audience but still producing digital data.

If in any doubt about what sharenting looks like, let me illustrate. I just did it, at the beginning of this chapter. I did it in Chapter Four as well. And I'm about to do it some more.

Having expatriated myself some decades ago, and living far away from many family members and hometown friends, I always found social networks invaluable for maintaining social connections in the country of my birth. As long as my privacy settings were carefully calibrated, and my posts mindfully made, I considered the benefits of Facebook to outweigh the drawbacks. True to the above-quoted statistics, I was a sharenter from the first sonogram, and my daughter certainly did have a substantial digital footprint before the age of two. That age feels particularly significant, in fact, because it represents a point at which I exponentially augmented her online presence. Prior to the age of two, I had primarily posted her images, but once she started to hold proper conversations, the Zoe Dialogues began.

The Zoe Dialogues consisted of exact transcriptions of conversations between myself and my daughter, on a variety of topics. Over the course of many years, my online community came to know her thoughts on art, religion, dinosaurs, chicken farming, table etiquette, nature and the Internet. She tackled the big existential questions – memory, finitude, responsibility, God, love and death. Particularly popular amongst my friends, and always guaranteed to notch up the number of likes and comments, were her wry observations about the behaviour and preoccupations of adults: drinking, working, parenting. Also faithfully transcribed were her commentaries on musical genres and the musicians who contributed to them. Blues, bluegrass, punk, classic rock and roll, alternative rock. The Ramones, Frank Zappa, ZZ Top, Lynyrd Skynyrd, the Dandy Warhols, Pulp, David Bowie.

Our little muso, her father and I thought, impressed, loving the fact that she was so much like us. How much these dialogues always brighten up my day, my friends said. Like, like, love, love, love, like – little hits of positive reinforcement, little bits of incentive to keep at it. And why not? My audience was eager, my privacy settings rigorously maintained, and I loved having this record of her childhood, so much more comprehensive than the few lines written by my mother in my own baby book.

After a while, though, I became anxious about the same thing that had bedevilled Greg. What if something happened to Facebook? Shouldn't I back up all of these precious conversations somehow? It's an understandable concern. In July of 2018, Facebook's stock dropped dramatically, the site losing $120 billion in value, a stark reminder that no company stays on top for ever.[336] And so, on my daughter's seventh Christmas, I copied and pasted all of the Zoe Dialogues into one document and printed it up as a physical book, complete with a hard cover, ISBN code, and captions for each dialogue, as inspired by Michel de Montaigne's *Essays*. 'On language', 'On beauty', 'On friendship'. In the process, I discovered that over the preceding five years, I had posted so many dialogues on the social network that it ran to 175 pages of A4.

I gave a copy of *The Tao of Zoe* to one of her grandmothers, who didn't have social media and hence had never seen any of this. She was shocked and delighted that I'd recorded so much, so faithfully. Isn't it wonderful, everyone said, that you are capturing these memories of her childhood! Yes, that's what I'm doing, I thought, capturing memories, and it *is* wonderful. I now realise that it was far more than 'capturing'. I'm not a mirror. I'm a prism. I'm the photographer holding the camera, the teller of the tale. I'm making my decisions and selecting my moments. I'm a narrator of my version of her story, as much writing the story as capturing it.

I remember the precise moment in time that I awakened to the extent of my power. My daughter was six years old, and we had gone to dinner with people who were long-time friends of mine but who had never met my daughter face to face. They were beside themselves. 'The famous Zoe! I can't believe I'm finally meeting you! I feel like I know you!' they gushed. They waited with anticipation for her to drop her usual wisecracks and hilarious insights, of which they'd read so many, into the conversation. When in due course she delivered the goods as expected, they were thrilled, overflowing with positive reinforcement. 'Oh my gosh, it's true, it's really true, you're just like I thought you'd be!' they enthused appreciatively. Grilling her on her musical favourites, they anticipated what she'd say before the answers were out of her mouth. 'You've got such cool taste in music!' they said, expectantly. If she named a current pop song, part of the soundtrack of the most recent Pixar film perhaps, their tone and faces let her know that she'd let them down somewhat. 'Really?' they probed, hoping for a retraction, or at least a more satisfactory answer to follow. 'That's not what I would have expected you to say. Aren't you more of a Bowie girl?'

On the way home there was silence for a while from the car seat, and I thought she was asleep. But the voice that eventually issued from the back was fully awake. 'I don't understand,' she said, not sounding entirely pleased. 'How do these people know me? Am I famous or something?' My daughter is hardly famous, but thanks to her besotted mother, she has a celebrity's awareness of what it is like to be on the receiving end of a parasocial relationship, to have little awareness or control of who knows what about her – maybe this is what was on Bowie's mind as he sang, in 'Lazarus', *everybody knows me now*.[337] My transcriptions might have been utterly exact, no more or less than what had actually issued from our lips, but was my role as a biographer as neutral or beneficial as I had imagined?

The academics and media commentators who dissect and critique modern-day sharenting practices are concerned about a lot of things. What if paedophiles obtain and misuse a child's online image? What if the child's safety is compromised through a combination of geo-location technologies and the information about their routines and personally identifiable information that their trusting or unwitting parents post? What if a child's digital footprint disadvantages or harms them in later life – for example, how might a child be hurt, victimised or exploited – now or in the future – by electronic surveillance and targeted marketing? All of that is food for thought, but my primary concern is even more overarching than that.

Throughout the formative early years of my daughter's life, before she went to school and while we still largely controlled whom she met, nearly everyone she encountered was chock-full of pre-existing expectations of who she would turn out to be. She met virtually no one 'new', no one uninitiated, no one who did not have preconceptions about her, who had not already formed some degree of parasocial relationship with my child. Everyone 'knew' her, and they let her know it. Their expectations had been shaped entirely by the digital footprint created and controlled not by her, but by her mother, and endlessly reinforced by that mother's online social circle.

If you were surprised at the fact that a six year old apparently independently selected a David-Bowie theme for her birthday celebration, it probably makes a whole lot more sense to you now. Her choice – which seemed entirely spontaneous and volitional – was, in reality, shaped by *my* preferences. All this I have done. Because of social networking, I have been able to wield a far greater degree of parental influence than I would have had in a pre-digital era. However long Zoe's life may be, I suspect that through writing and sharing a narrative of my daughter online, from her very beginnings, I have already exerted considerable control over what her enduring legacy will be when she's no longer on earth.

Before that point comes, though, who can yet tell whether the digital die I have cast during my daughter's formative years will result in good or ill for her future life? The great-granddaughter of Elizabeth and James is just one of many, part of a fascinating generation. These young people were born into a revolutionary time, when we can barely understand the rapid sea changes taking place in our information-saturated society. The consequences of coming into being at that juncture in history are not yet known. Children take this strange and contradictory existence for granted.

And what is so contradictory about it? We seem to have all the power and control in the world over our information, and at the same time no power or control at all. Our digital information could either be indelible, affecting how others see us and how people will remember us for evermore, or it could vanish, eliminating any evidence that we ever existed and rendering us victims of a twenty-first century Dark Age. And while the era of information might have dangled the tantalising prospect of perpetuity in our faces, death has always found ways of defeating our fantasies of immortality. Everlasting life has always been an illusion, a mirage in the desert, and the digital age may not change that as much as one might expect.

This started out being a book about death in the digital age but, as it happens, death has mostly been an uncannily useful tool to better understand the transformations and challenges of how we live now. Viewed through the lenses of death and the digital, so much about life in these times looks clearer. Power, corporate control, ownership and privacy. Identity, freedom and choice. Connection, memory and legacy. Love.

Ω

Anticipating my upcoming marriage to Zoe's father, I went to visit a jewellery designer whose work I knew and loved, Sarie.[338] Her style is modern, organic and fluid, inspired by nature. She turned my grandmother's wedding and engagement rings over in her hands, inspecting the sparkling stones in their delicate, linear, old-fashioned setting. This particular ring belonged not to Elizabeth, but to my paternal grandmother Eva, a woman who was as warm and cuddly as Elizabeth was cool and flinty, but who was notoriously unsentimental about memorabilia. She was renowned for writing identifying names directly on the faces of people in photographs, and had never been completely forgiven for throwing away Super-8 footage of my father as a child.

Sitting in Sarie's workshop, I suddenly remembered sitting on my grandmother Eva's lap, twisting the ring around and around on her finger. Hit with a powerful sensory memory, I could vividly recall the warmth of her closeness, the bones of her knuckles, the fragility and near-transparency of her skin. Imagining her ring being disassembled, the diamonds plucked from their settings and the gold melted down, I felt a bit strange. Sarie, seeing my hesitation, looked at me with kind eyes. It's hard, she said. I wasn't so sure I wanted to do this any more. But then, looking at the colourful gemstones scattered about Sarie's worktable, it occurred to me: perhaps I could combine my grandmother's diamonds with my daughter's birthstone. But although my uncertainty diminished significantly at this thought, it did not disappear.

Sarie pointed out that with the design we were considering, there would likely be stones left over, a few tiny diamonds. 'Why don't I make a little companion ring for your daughter?' she suggested. 'With stones from her great-grandmother's ring? Your daughter is so much a part of the day.'

I thought of all the possessions that my grandparents had passed down, and how much I cherished them. I thought of the handwritten notes that Elizabeth had placed with various items, fascinating and funny and informative bits of family history. I gazed at Eva's ring, gleaming in Sarie's hand. I felt myself embodying the memories of these my ancestors, this my family. So many dear people, so many precious things, so many lovely memories, none of whom, and none of which, had ever been digitally captured or reflected or stored.

Yes, I said. Actually, I would love you to make a ring for me and for my daughter. Let's use these heritage materials to create something new, something perfectly suited to *this* moment – old and new, continuity and change intertwined. But in that same instant I found myself wiping away tears, tears triggered by memory, and by love, and by my awareness of the passing of all things.

Sarie understood, and for a moment we sat together silently. Sunshine streamed in through the window, across her antique workbench, warm on our faces and glinting off the surfaces of what was about to be transformed.

'What a lovely legacy,' Sarie said.[339]

The author's mother and daughter at the memorial for
James and Elizabeth

Last Words
A Decalogue for Your Digital Dust

Earth to earth, ashes to ashes, dust to dust. I'd been dry-eyed, but at that point I could feel a lump rising in my throat. It was one of my first funerals, and I'd only heard the phrase 'ashes to ashes' in films or song lyrics. I remembered a creation story from Sunday school, a tale of God scooping up a handful of earth and using his divine, life-giving breath to animate it into the first human, and maybe the Book of Genesis wasn't too far off – scientists too have said that we're all made up of the detritus shed from heavenly bodies. Tears blurred my vision. How odd it was to think of my complex, strong, gifted grandmother as a mere collection of stardust, her human spark now extinguished, her distinctiveness melting into the earth. The swan grave marker would eventually come to symbolise her singularity, but that sits above the ground. Under the ground, she's now no different to her cemetery neighbours, slowly breaking down into indistinguishable atoms.

I know that one day the great leveller will visit me too, and my physical body will be reduced to dust, just like my grandmother's. Unlike her, however, I'll also knowingly leave behind digital dust, the remainder of decades of online cell-shedding. Unless you're living off the grid as a technology-averse hermit, so will you. Your digital dust, floating in the ether for an indeterminate period of time, won't require a god to gather it up and imbue it with a kind of vitality. Anyone with an Internet connection will be able to do that, although they'll be searching for different things – if not a spark of actual life, perhaps a glimmer of something else. Those who knew you in life may seek solace, memories, answers, a sense of connection. Your descendants, driven by curiosity and personal meaning-making, may discover you as they explore where – and who – they came from. Perfect strangers may learn from the contributions you've made to the world or may find inspiration in your life, and criminals may cloak themselves in your posthumously persistent digital identity, impersonating you for nefarious purposes. What should all this mean to you, and what should you do about it?

In my job as a psychologist, I explore problems with my clients and help them find their own answers. Accustomed as I am to letting people form their own conclusions, dispensing advice or being overly directive goes slightly against the professional grain. On the other hand, I've spent more than a decade talking and thinking about death in the digital age, and I'd be lying if I said I hadn't formed any opinions during that time. My vantage point is framed by my personal circumstances and history, and your perspective is bound to be different. Nevertheless, at the end of this book's journey and for what it's worth, I offer you ten general principles. At most, take them as a set of commandments for life and death in our age, a decalogue to help you constructively confront the realities of digital dust. At least, take them as food for thought.

I

Confront your death anxiety. Facing up to your eventual physical demise can be a struggle, understandably. Death is the great mystery, the ultimate loss of control, the unfathomable snuffing out of the conscious existence you've always known. You may tend to avoid the things that make you anxious, but avoidance can maintain and fuel anxiety. Take a deep breath and look your finitude square in the eye, and you may find that you can bend this awareness to your advantage, in service of a more conscious life. Occasionally remind yourself that each time you use a connected device, you may be laying another brick on your eventual digital monument, writing another paragraph in an autobiography that will outlive you. If you see digital activity as not just the stuff of *life* but the stuff of *legacy*, you'll probably make smarter decisions. If you died tomorrow, would you be content with what's left behind? Would you be happy with the words that hover in the ether, with the image that lingers in your digital reflection? If not, what do you want to do differently, online and off?

II

Always assess, never assume. This rule applies whether you're planning for your own digital legacy or managing someone else's. Remember that what was true in the pre-digital era may not, and often does not, apply in digital contexts. For all of the online accounts that matter to you, look for the terms and conditions that explain what happens to your data when you die. If you can't find them, ring up and enquire – maybe your questions will spark some change where it's needed. Get education and advice from online organisations that educate the public about digital legacy, and if you're choosing a probate specialist, try to find one who's knowledgeable about digital assets.

III

Put yourself in other people's shoes. First, if you are logged on to, managing, or actively operating a deceased person's phone, social media, blog, message apps and/or email accounts, it's very important to realise that any activity from within these accounts can result in 'voice from beyond the grave' phenomena. Both academic research and anecdotal evidence indicates that this will likely have a shocking and negative impact on others. You'll also be accessing data associated with still-living persons, which might certainly concern them and which, on reflection, might also concern you. Second, maybe you don't care what happens after you die, so you don't concern yourself with your digital remains. Maybe you want your online presence deleted after you draw your last breath, and that's your prerogative. Before you make any decisions, though, see it from the other person's perspective. If someone dear to you had a rich online presence, replete with photos, videos and other memorabilia, would it bother you if all that material disappeared after their death? Let that sink in, really sink in, and you may suddenly feel less devil-may-care about your own digital legacy. Using your imagination only takes you so far, though, and you may be making erroneous assumptions about how others would feel. (That brings us to the fourth rule.)

IV

Have conversations about death and the digital. Talk about it because it's an intellectually interesting topic, by all means, but don't keep it in the abstract – meaningfully apply it to your own life. Speak to the people you love about how you feel about your digital material, about your personal take on privacy and ownership of data in the digital era, and about what you'd like to happen with your data when you're no longer here. Ask them about their thoughts and preferences regarding their own digital information. If your friend says she wants all of her social media deleted on death and you think it would break your heart to lose it, tell her so, while also trying to understand where she's coming from. (If she sticks to her guns, think about how you might preserve anything that's precious to you.) If it's important to you to have your Facebook profile memorialised or your blog maintained for posterity, make that clear to your next of kin, who may otherwise have had no idea. If you're not sure how you feel, talking it through with others will often clarify your position considerably.

Furthermore, don't just have conversations about this topic in the social realm. Think about your workplace, if you have one. Does your employer control or process the data of employees, clients, users and customers? Could any of this information be of practical, financial or sentimental value to anyone else, if the persons associated with the data were to die? Does the organisation you work for have well-articulated policies and procedures around what to do with the data of the deceased, and clear guidance and training on how to interact with bereaved individuals? The answer to all these questions may well be no, because we still have a long way to go in this area. Lead the change by speaking up.

V

Make a digital-era will. First, buck the trend by making a will at all. Then, flout convention even further by explicitly recording your wishes for digitally stored material. Ironically, this thoroughly modern last will and testament may need to be on paper, signed in ink, and officially witnessed – not all jurisdictions accept digitally generated and stored wills – so check the law in your land. Even if digital 'things' aren't legally executable where you live, it never hurts to be clear about what you want. For example, 'I would like my mother to use money from my estate to keep up payments on my blog' or, 'I want my Facebook profile to be memorialised'. In the absence of a clear steer from you, both your loved ones and the corporations that manage your data will make inferences – and potentially argue amongst themselves – about what you would have wanted. Maybe they'll get it right. Maybe they'll get it wrong.

On every account that allows you to do so, nominate a digital executor (Legacy Contact on Facebook and Inactive Account Manager on Google, for example), making sure you've discussed this with the person(s) first. Again, this might not hold legal weight where you live (yet), but it's far better than giving no indication at all of your desires. Repeat yourself in your official will or in a letter kept with your official will. Remember, though, that you can't bequeath what you don't own, and you might not be allowed to transfer control of your account or give access to its contents to another individual in a one-account-one-user situation. Manage your expectations by always checking the T&Cs.

VI

Develop a master-password system that's accessible by a trusted person. Password-protecting your devices, accounts and apps is absolutely necessary to maintain your data security and privacy in life, and we're encouraged to change them periodically or when there's been a breach. After death, these eminently sensible safeguards can become a nightmare for those who need to settle your estate or access personal memorabilia. Since our passwords tend to be numerous and changeable, it can prove impractical to maintain an up-to-date list of passwords specifically for legacy or estate-planning purposes. If you use a *well-established* online password vault or allow your phone to store passwords in its settings, you can lodge and update just the master password for the service or phone somewhere secure – in your will, a sealed envelope in a safe, with your solicitor, or with trusted other(s). If your phone has biometric features, you could even give your solicitor and/or another trusted person fingerprint or face-recognition access – although keep in mind that if the battery is discharged on the phone is powered off, a password might then be required again.

A few caveats, though. First, if someone allows you in-case-of-death-or-emergency access to their devices and account passwords, don't abuse their trust by snooping. For the sake of whatever relationship you have, be clear between yourselves about the circumstances under which it's permissible to access the other's private data. Second, remember that account passwords will have a shelf life if terms and conditions stipulate account deletion, memorialisation or some other type of login lockdown upon the confirmed death of the original account holder. Finally, if you have skeletons in your digital closet that you suspect would cause confusion, feelings of betrayal or additional pain to people you care about, be careful about granting those individuals carte blanche access. If you're unwilling to confess and address any lurking secrets while you're still alive, you might wish to embark on a deleting frenzy, let the passwords die with you, or otherwise ensure that the closet door stays shut when you're no longer able to guard it. In some circumstances – including ones where you won't be around to pick up the pieces – there's an argument for allowing your dear ones to remain blissful in their ignorance.

VII

Be an unapologetic curator. 'Curating' your online presence has received a bad rap. We associate it with people artificially inflating the amount of excitement, happiness and success they're enjoying and showing off airbrushed lives that provoke insecurity and envy in others. From a legacy perspective, though, there's a lot to be said for curation. I'm not talking about the censorship of difficult stuff here – I'd argue that it's quite good both for you and for others to disclose tough stuff as well as good stuff. Instead, I'm referring to selectivity that's in service of manageability.

Imagine inheriting the laptop or other storage devices of someone who took a *lot* of photographs. Naturally, because of the ubiquity of digital photography and in-phone cameras, this describes a lot of people. You'd be confronted with thousands of images. Beyond those files there might be thousands of other types of files (documents, emails, text messages), most of them without any context or comment to indicate what's meaningful or important and what's not. Many bereaved people find themselves so staggered by the huge, undifferentiated mass of a loved one's digital stuff that they can't see the wood for the trees and avoid engaging with it. An ardent genealogist of today may go into raptures when they discover a single photograph of their great-great-great-grandfather, from a time when photographic images were rare rather than abundant. If all of today's digitally stored data manages to be persistent and accessible over the next centuries, amateur genealogists will find a once-challenging hobby about as stimulating as shooting fish in a barrel.

On our curated social media platforms, we're selective. We present the things that mean the most to us, that move us the most, or that represent us the way we want to be seen. Unlike the mass of files on your computer, or the results from an online search of your name, a social media profile offers context, timeline and narrative – as such, it's far easier for someone to metabolise, recognise and maybe even experience it as 'you'. Yes, it's a heavily edited and maybe even skewed reflection of you. But for memory and legacy purposes, so what if your profile selectively emphasises the happiest times, the most exciting experiences, and the moments when you looked particularly fetching? Whether you're using your digital autobiography to reminisce in later years, or whether your loved ones or descendants are accessing it to remember or discover you, perhaps a rosy tint is no bad thing.

VIII

Accessing more doesn't always equate to feeling better. In grief, it's common to 'search' for the lost loved one and to hungrily fall upon anything associated with them. Particularly in the case of difficult deaths – deaths that were unexpected or traumatic, or that involved suicide, homicide or mystery – this searching can assume the form of combing through the deceased's data in a relentless quest for answers, comfort and resolution. I'm sure that in certain circumstances going through private correspondence can be helpful. Often, however, the searcher will encounter material that is frustratingly ambiguous, or painful, or that only provokes more uncertainty. The dead person is no longer there to answer unsettling questions, to resolve ambiguities, to provide important context, or perhaps to apologise and seek resolution or healing. This is one of the aspects of the information age that can make grief particularly harrowing. When you delve into a deceased loved one's digital data, you could be opening a precious gift or Pandora's box. Which one it will be is impossible to discern ahead of time, so be careful.

IX

Go old school. Papyrus scrolls from ancient Egypt are still with us, but our digitally stored information may be nonexistent, inaccessible and/or unreadable in decades or centuries, much less millennia. We just don't know. What we can say with comparative certainty, though, is that paper and other material objects may be vulnerable to the elements, but they won't fall victim to technological obsolescence. Even if you have confidence that your digital dust will stick around a good long while, and even if you think that you've ensured that others will be able to access it, things change. Technology evolves and obsolesces in predictably unpredictable ways. Given that, it's worth selectively transforming digital stuff into material stuff now and again. Make the occasional photo book of the images that you love the most or that commemorate meaningful occasions. Print out the electronic correspondence that's closest to your heart. Transcribe a passage of your online writing into a beautiful notebook that goes on a physical shelf.

If you never get around to any of that, don't despair. The digital world hasn't destroyed your ability to connect with your loved ones and your ancestors in the same ways people always have. My memories of my grandparents are triggered by photographs and letters, but they may also bubble up when I see or touch or use a certain ring, an old button in the sewing box, or a dog-eared and yellowing recipe card giving instructions for chocolate chip cookies. They are reinforced and expanded in the mutual sharing of recollections with others who knew them. I hold their stories in my heart and mind, and I pass them on to the next generation. The more things change, the more they stay the same.

X

Forget immortality. For a handful of us, like Da Vinci, Shakespeare, Bach or Hawking, we will make contributions to art, literature, music or science that ensure our fame for many centuries. For most of us, though, the digital and material remains of our lives will be primarily important to those who knew and loved us, to this generation and the next. Knowing this, focus on living the life you want and doing the things you value. Ask yourself this question, often: even if no one ever remembered me for this, even if this action never led to fortune or lasting fame, would it still be worth doing? Would it still be good for me or someone else?

Maybe fantasies of immortality are fuelled by a dread of death and insignificance. Maybe they're associated with an inflated sense of self-importance. Maybe they're driven by the desire for your life's accomplishments to make a difference in the world for as long as possible, and why not? Whatever a quest for immortality is about, it's a losing battle, and what might you miss out on while you're fighting it? By all means, capture, record and share the meaningful things and the good work of your life. I have no doubt that my daughter and her descendants will enjoy the lovingly crafted account of her childhood that I've built over the years on social media. Through those efforts, my (and her) digital dust may well bring some measure of joy, interest and even inspiration to future generations, and I'm still pleased about the time and attention I've devoted to accumulating it. Ultimately, however – in both importance and lasting impact – it will pale in significance to the amount of time and attention I devoted *to her.*

The essence of this last rule, therefore, is this. Live the best, most valued life you can. Love well. Be present in and grateful for the moment as it's lived. Be a force for good in the world. Devote yourself to all that, and a legacy of which you can be proud – whatever form it assumes, and however long it lasts – will take care of itself.

Acknowledgements

This book would not have been possible without the extraordinarily generous and loving support and input of many people. Most instrumental in ensuring that it saw the light of day were my brilliant literary agent, Caroline Hardman of Hardman & Swainson; the talented novelist Rosie Fiore-Burt, without whose introduction I would never have met Caroline; and Andrew McAleer, my fantastic editor at Little, Brown UK, who 'got' this book from the beginning and has been a staunch cheerleader for it throughout. The lovely folks at Urban Writers Retreat, This Time Next Year London, All You Read is Love, and Blacks Club always provided me with the spaces I needed to get the job done. A literally uncountable number of people in London – particularly my wonderful Leytonstone community – were always eager to hear about the book and encouraged me when I wavered; particular thanks go to Abi and Dave Hopper and Michael Nabavian, the latter of whom provided an extensive critique and commentary on the original proposal in his usual thorough way.

Endless thanks to all who helped with the research for this book, whether through discussion, inspiration or direct contribution. They include Nick Gazzard, Kate Roberts, Edina Harbinja, Lilian Edwards, Tony Walter, Debra Bassett, Morna O'Connor, Dennis Klass, Vered Shavit, Paula Kiel, Marc Saner, Jane Harris, Sinead McQuillan, Jed Brubaker, Lisa Jones McWhirter, David Costello, John Troyer, Edith Steffen, Stacey Pitsillides, Caroline Lloyd, Jennifer Roberts, Gary Rycroft, James Norris, Lucy Watts, Beth Eastwood, Helen Holbrook,

Pete Billingham, Sharon Duffield, Amy Harris, Susan Furniss, Jon Reece, Gary Burks, Marius Ursache, Peter Barrett, Steve McIlroy, Greg King, Kristin Furnish, Sarie Miell, Wendy Moncur, Maggi Savin-Baden, Silke Arnold-de-Simine, Carl Öhman, David Watson, the members of the Death Online Research Network, and numerous anonymous research participants. You have all been so generous with your time and with your experiences. Those who have departed this life could not give their consent for their very personal involvements in this project, and I am grateful to them too.

And last but by no means least, my family. Thank you particularly to my parents, including my mother Beth, daughter of James and Elizabeth, who championed (and attempted to edit) my earliest authorly efforts as a child and whose story provided me with a 'way in'. My husband Marcus endured endless hand-wringing about the writing process with patience and good grace, and my mother-in-law Jan went above and beyond with childcare so that I could finish on time. Eternal gratitude to my perpetually inspiring daughter, Zoe, for her love and her loveliness and her direct contribution to this book, as well as for the wise perspectives that she has expressed over the years on life, death, afterlife and the online environment. Finally, I owe so much to James Fisher, Elizabeth Fisher and Eva Rudwell – this is for you as much as it is from you, in love, and in memoriam.

Endnotes

Introduction: Remembering Elizabeth

1   Image © Elaine Kasket 2017.

2   Image © Elaine Kasket 2017.

3   There has been much written about the terms 'digital immigrant' and 'digital native', and some papers attribute particular qualities to these respective categories. For example, it has been claimed that digital immigrants prefer to talk in person, are logical learners, prefer to do one task at a time; digital natives are always online, intuitive learners, are multitaskers. These descriptors/ characteristics are contested. Here, I am simply using the immigrant/native terminology to refer to people's temporal relationship with the digital age. See Prensky, M., 'Digital Natives, Digital Immigrants'. *On the Horizon*, 9/5 (2001), 1–6, retrieved 14 July 2018 from http://www.marcprensky.com/writing/Prensky%20-%20Digital%20Natives,%20Digital%20Immigrants%20-%20Part1.pdf

4   This phrase echoes the title of a book by Paul Kalanithi, a young neurosurgeon facing his own death who wrote about what makes a life worth living. Kalanithi, P., *When Breath Becomes Air* (New York: Penguin Random House, 2016).

Chapter One: The New Elysium

5   I was asked to participate in this event by the artist Gillian Steel. You can read about the event in Alistair Quietsch's review of the exhibition, available on http://www.aestheticamagazine.com/what-happens-when-we-die-stardust-some-thoughts-on-death-at-st-mungos-museum-glasgow/, last retrieved 27 July 2018.

6   I wrote about my findings in an academic research article, authored shortly after this event. See Kasket, E. 'Continuing Bonds in the Age of Social Networking: Facebook As a Modern-Day Medium', *Bereavement Care*, 31/2 (2012), 62–9. Available on: https://www.tandfonline.com/doi/abs/10.1080/02682621.2012.710493

7   A comprehensive text explaining and exploring the concept of continuing bonds is Klass, D., & Steffen, E., *Continuing Bonds in Bereavement: New Directions for Research and Practice* (New York: Routledge, 2017). This is the updated text from the original, 1996 *Continuing Bonds* book, edited by Dennis Klass, Phyllis Silverman, and Steven Nickman.

8   For a print version of Freud's essay, see Freud, S., 'Mourning and Melancholia', in J. Strachey (ed. and trans.), *The Standard Edition of the Complete Psychological Works of Sigmund Freud*, vol. 14 (London: Hogarth Press, 1917), 252–68. I accessed it from http://www.arch.mcgill.ca/prof/bressani/arch653/winter2010/Freud_Mourningandmelan-cholia.pdf

9   Segrave, K., *Women Swindlers in America, 1860–1920* (London: McFarland & Company, 2007), p. 7.

10 Ibid.

11 Lescarboura, A. C., 'Edison's Views on Life and Death', *Scientific American*, 123/18 (1920).

12 I'm thinking of course of the 1982 film *Poltergeist*, written and produced by Steven Spielberg and directed by Tobe Hooper.

13 For a moving account of the wind telephone, listen to the *This American Life* episode of 'One Last Thing Before I Go'. Available on https://www.thisamericanlife.org/597/one-last-thing-before-i-go, accessed 20 July 2018.

14 Elysium, also known as the Elysian Fields, is one location for the afterlife that is described in Greek classical literature. Akin to the Christian notions of Heaven that came later, Elysium was reserved for relatives of the gods as well as those who had been particularly good, righteous and heroic during life.

15 This exhibition ran from 24 October 2015 to 13 March 2016. I appeared on the *BBC Breakfast* TV show speaking about death and the digital on 4 December 2015, broadcast live from the museum. Link for viewing not available. You can see a description of the exhibition on https://www.bristolmuseums.org.uk/bristol-museum-and-art-gallery/whats-on/death-human-experience/ – last accessed 28 July 2018.

16 Quote recorded in personal journal, 5 December 2015.

17 Pontin, J., 'Silicon Valley's Immortalists Will Help Us All Stay Healthy', *Wired* (15 December 2017). Available on https://www.wired.com/story/silicon-valleys-immortalists-will-help-us-all-stay-healthy/, accessed 27 July 2018.

18 From 'Raising Children in the Early 17th Century: Demographics', a monograph compiled by Plimoth Plantation and the New England Historic Genealogical Society, n.d. Available on https://www.plimoth.org/sites/default/files/media/pdf/edmaterials_demographics.pdf, accessed 27 July 2018.

19 The World: Life Expectancy (2018), available on http://www.geoba.se/population.php?pc=world&type=015&year=2018&st=country&asde=&page=3, accessed 27 July 2018.

20 'Demographia: Greater London, Inner London & Outer London Population & Density History'. Available on http://www.demographia.com/dm-lon31.htm, accessed 27 July 2018.

21 For a history of the Magnificent Seven, see Turpin, J., & Knight, D., *The Magnificent Seven: London's First Landscaped Cemeteries* (Stroud, Gloucestershire: Amberley Publishing, 2011).

22 See Arnold, C., *Necropolis: London and Its Dead* (London: Simon & Schuster UK, 2008).

23 Greenfield, R., 'Our First Public Parks: The Forgotten History of Cemeteries', *The Atlantic* (16 March 2011). Available on https://www.theatlantic.com/national/archive/2011/03/our-first-public-parks-the-forgotten-history-of-cemeteries/71818/, retrieved 27 July 2018.

24 See Rothman, L., '*TIME* Has Been Picking a Person of the Year Since 1927. Here's How It All Started', TIME.com (16 December 2017). Available on http://time.com/5047813/person-of-the-year-history/, retrieved 27 July 2018.

25 Photo illustration for *TIME* by Arthur Hochstein, with photographs by Spencer Jones/Glasshouse.

26 Amira, D., 'The One Joke You Should Never Make in Your Twitter Bio', *New York Magazine – Daily Intelligencer* (20 September 2013). Available on http://nymag.com/daily/intelligencer/2013/09/twitter-bio-time-person-of-year-man-2006.html, retrieved 27 July 2018.

27 Aced, C., 'Web 2.0: The Origin of the Word That Has Changed the Way We Understand Public Relations', International PR Conference: Images of Public Relations (2013).

28 DiNucci, D., 'Fragmented Future', *Print* (1999). Available in full text on http://www.darcyd.com/fragmented_future.pdf, retrieved 27 July 2018.

29 Many bloggers referred to this statistic. One such was Sifry, D., 'The State of the Blogosphere' [personal blog] (6 August 2006). Available on http://www.sifry.com/alerts/2006/08/state-of-the-blogosphere-august-2006, retrieved 14 July 2018.

30 Grossman, L., 'Person of the Year 2010: Mark Zuckerberg', TIME.com (15 December 2010). Available on http://content.time.com/time/specials/packages/article/0,28804,2036683_2037183_2037185,00.html, accessed 27 July 2018.

31 Smith, A., 'Record Shares of Americans Now Own Smartphones, Have Home Broadband', Pew Research Center (12 January 2017). Available on http://www.pewresearch.org/fact-tank/2017/01/12/evolution-of-technology/, retrieved 14 July 2018.

32 For an example of such satire, have a look at Audritt, D., & Butterfield, K., 'Anti Social – A Modern Dating Horror Story' [video], Comic Relief Originals (2017). Available on https://www.youtube.com/watch?v=GEWnXmD-fVZg, accessed 27 July 2018.

33 These categories are each useful in some way for thinking through the themes in this book, but they are informally rather than scientifically arrived at and hence are neither mutually exclusive nor exhaustive.

34 Lunny, O., 'SMS Is 25 Years Old – And Still Going Strong', ITProPortal (7 December 2017), available on https://www.itproportal.com/features/sms-is-25-years-old-and-still-going-strong/, retrieved 14 July 2018.

35 Plenty of articles document the 'free fall' in the camera market that has occurred since – and that is attributable to the rise of smartphones. One such article, with a helpful graphic, is Djudjic, D., 'Camera Sales Report for 2016: Lowest Sales Ever on DSLRs And Mirrorless', DIY Photography (2 March 2017). Available on https://www.diyphotography.net/camera-sales-report-2016-lowest-sales-ever-dslrs-mirrorless/, accessed 27 July 2018.

36 For an accessible discussion of context collapse online, see this blog post by sociologist Michael Wesch of Kansas State University. Wesch, M., 'Context Collapse' [blog post], Digital Ethnography [blog] (2008). Available on http://mediatedcultures.net/youtube/context-collapse/, accessed 27 July 2018.

37 Anderson, J. Q., 'Millennials Will Benefit and Suffer Due to Their Hyperconnected Lives', Pew Research Center: Pew Internet & American Life Project (29 February 2012). Available on http://www.pewinternet.org/files/old-media/Files/Reports/2012/PIP_Future_of_Internet_2012_Young_brains_PDF.pdf, retrieved 14 July 2018.

38 BBC News, 'Shutter Falls on Life-Logging Camera Startup Narrative' (28 September 2016), available on https://www.bbc.co.uk/news/technology-37497900, retrieved 14 July 2018.

39 Eggers, D., The Circle (New York: Knopf, 2013).

40 Ponsoldt, J., & Eggers, D., The Circle [film] (2017). A Netflix Original film. These two screenwriters adapted Dave Eggers's 2013 dystopian novel for Netflix.

41 Graphic © Elaine Kasket 2018. This represents only one of many ways the different elements of our digital footprint could be classified, a useful categorisation for the purposes of this book.

42 Eastwood, B. (producer), 'We Need to Talk about Death: Digital Legacy' [radio programme], BBC Radio 4 (2017). UK

listeners can hear the episode – which is Series 2, Episode 2 of 'We Need to Talk about Death' – on https://www.bbc.co.uk/programmes/b09kgksn

43 The website for the Digital Legacy Association is https://digitallegacyassociation.org – more will be said about the DLA in this book. For full disclosure, although my involvement at time of writing has been limited, I am the Bereavement Lead for the DLA.

44 Again, this worksheet is one of many and gives a more or less representative indication of the scale of digital assets that need consideration. See https://njaes.rutgers.edu/money/pdfs/Digital-Assets-Worksheet.pdf, accessed 14 July 2018.

45 A YouGov survey in 2017 in the UK reported that fewer than 4 in 10 adults had made a will. See https://reports.yougov.com/reportaction/willsandprobate_17/Marketing. Similar percentages hold in the United States. A 2016 Gallup poll survey reported that only 44 per cent of American adults have made a will. See https://news.gallup.com/poll/191651/majority-not.aspx, both surveys accessed 27 July 2018.

46 The names of all actors in this story have been changed. Story based on audio-recorded, fully transcribed interview with daughter 'Kim'.

47 Although I didn't formally interview John in a recorded conversation for this book, he and I did speak on several occasions throughout the writing of this book – this conversation took place at a pre-performance workshop for the immersion theatre piece 'User Not Found', at the University of Reading.

48 Tow, C. S., 'Timeline: Now Available Worldwide' [Facebook post] (24 January 2012). Available on https://www.facebook.com/notes/facebook/timeline-now-available-worldwide/10150408488962131/, accessed 28 July 2018.

49 Callison-Burch, V., Probst, J., & Govea, M., 'Adding a Legacy Contact' [Facebook Newsroom] (12 February 2015). Available on https://newsroom.fb.com/news/2015/02/adding-a-legacy-contact/, accessed 28 July 2018.

50 This Day in Quotes (2 June 2018). Reports of Mark Twain's quip about his death are greatly misquoted. Available on http://www.thisdayinquotes.com/2010/06/reports-of-my-death-are-greatly.html, accessed 14 July 2018.

51 Abrams, R., 'Google Thinks I'm Dead. (I Know Otherwise)', *New York Times* (16 December 2017). Available on https://www.nytimes.com/2017/12/16/business/google-thinks-im-dead.html, accessed 14 July 2018.

52 For the World Wide Cemetery, see cemetery.org, accessed 28 July 2018.

53 Berreby, D., 'Click to Agree with What? No One Reads Terms of Service, Studies Confirm', *Guardian* (3 March 2017). Available on https://www.theguardian.com/technology/2017/mar/03/terms-of-service-online-contracts-fine-print, accessed 14 July 2018.

54 Olsen, S., 'Yahoo Releases E-Mail of Deceased Marine', C/Net (22 April 2005). Available on https://www.cnet.com/news/yahoo-releases-e-mail-of-deceased-marine/, accessed 28 July 2018.

55 This US act was passed in 1986, concurrent with the rise of electronic communications. For more information see https://www.epic.org/privacy/ecpa/, accessed 30 July 2018.

56 Most news outlets reported that John Ellsworth received just the CD, but there were these boxes of paper matter in addition, according to the official inventory. For a thorough description and a discussion of the legalities in this case, see Cummings, R. G., 'The Case Against Access to Decedents' Email: Password Protection As an Exercise of the Right to Destroy', *Minnesota Journal of Law, Science & Technology* (2014). Available on https://conservancy.umn.edu/bitstream/handle/11299/163821/Cummings.pdf, accessed 28 July 2018.

57 Ibid.

58 Personal representatives of the estate, in this case, John Ellsworth, Justin's next of kin.

59 Cummings (2014), 941–2.

60 Franco, M., 'How to Reclaim Your Digital Privacy from Online Tracking', *Lifehacker* (12 December 2017). Available on https://lifehacker.com/how-to-reclaim-your-digital-privacy-from-on-line-trackin-1820878546, accessed 28 July 2018.

61 Ibid.

62 See Geoff and Glenn in action on https://www.youtube.com/watch?v=Qg4-gDDx-As0, retrieved 14 July 2018.

63 This was all over the press, but for one article describing the overall situation, written a couple of weeks after it occurred, see *Wired* staff, 'AOL's $658 Million Privacy Breach?' *Wired* (7 August 2006). Available on https://www.wired.com/2006/08/aol_s_658_milli/, accessed 28 July 2018.

64 You can find the entire series of their work on YouTube. The first minimovie of *I Love Alaska* can be found on https://www.youtube.com/watch?v=c-SOCGdPyNU, retrieved 14 July 2018.

65 This is the descriptor for the minimovies on the Submarine Channel website, available on https://submarinechannel.com/i-love-alaska-new-minimovie-now-online/, accessed 28 July 2018.

66 Episode 8/13 of *I Love Alaska*, on https://www.youtube.com/watch?v=7iUHp-80vd0Y, accessed 14 July 2018.

67 Barbaro, M., & Zeller, T., 'A Face Is Exposed for AOL Searcher No. 4417749', *New York Times* (9 August 2006). Available on https://www.nytimes.com/2006/08/09/technology/09aol.html, accessed 14 July 2018.

68 Vered Shavit, personal communication, 20 December 2017.

69 Brannen, K., 'Her Secret History: I Discovered My Mother's Digital Life After Her Death', *Guardian* (8 May 2016). Available on https://www.theguardian.com/lifeandstyle/2016/may/08/secret-history-my-mothers-digital-life-after-her-death, accessed 14 July 2018.

Chapter Two: The Anatomy of Online Grief

70 You can view a video of my presentation at the October 2014 Irish Childhood Bereavement Network conference here: https://www.youtube.com/watch?v=f-08p9HmgBRU, accessed 28 July 2018.

71 Taken from Q&A notes at my Psychological Society of Ireland talk, Dublin, 3 October 2014.

72 I presented these research results at a Division of Counselling Psychology conference in 2013, underscoring the need for practitioners to think about how their assumptions and biases about the digital age might be affecting their work. Kasket, E., 'Grief on Social Networking Sites and Implications for Bereavement Counselling'. Individual paper presented at the Division of Counselling Psychology conference, Cardiff, Wales, July 2013.

73 Toolis, K., *My Father's Wake: How the Irish Teach Us to Live, Love and Die* (Boston: Da Capo Press, 2018).

74 Toolis, K., 'Why the Irish Get Death Right', *Guardian* (9 September 2017). Available on https://www.theguardian.com/lifeandstyle/2017/sep/09/why-the-irish-get-death-right, accessed 30 July 2018.

75 Ibid.

76 Morton, R., 'Facebook from Beyond the Grave', *The Weekend Australian* (25 February 2013). Available on: https://www.theaustralian.com.au/news/inquirer/facebook-from-beyond-the-grave/news-story/fdc31790d7a31c40bbc8f-c75246b2de8?sv=38a0139e-350807449c3822939be54596, accessed 28 July 2018.

77 Milovanovic, S., 'Man Avoids Jail in First Cyber Bullying Case', *The Age* (9 April 2010). Available on https://www.theage.com.au/national/victoria/man-avoids-jail-in-first-cyber-bullying-case-20100408-rv3v.html, accessed 28 July 2018.

78 Ibid.

79 Ibid.

80 BBC, 'Facebook Removes Memorial Page of Brazilian Journalist' (25 April 2013). Available on http://www.bbc.co.uk/news/world-latin-america-22299161, accessed 28 July 2018.

81 Pennington, N., 'Tie Strength and Time: Mourning on Social Networking Sites', *Journal of Broadcasting and Electronic Media*, 61/1 (7 March 2017), 11–23. DOI: 10.1080/08838151.2016.1273928

82 Rossetto, K. R., Lannutti, P. J., Strauman, E. C., 'Death of Facebook: Examining the Roles of Social Media Communication for the Bereaved', *Journal of Social and Personal Relationships*, 32/7 (21 October 2014), 974–94. DOI: 10.1177/0265407514555272

83 Kübler-Ross, E., *On Death and Dying* (New York, NY, US: Collier Books/Macmillan Publishing Co., 1970).

84 For a print version of Freud's essay, see Freud, S., 'Mourning and Melancholia', in J. Strachey (ed. and trans.), *The Standard Edition of the Complete Psychological Works of Sigmund Freud*, vol. 14 (London: Hogarth Press, 1917), 252–68. Available on http://www.arch.mcgill.ca/prof/bressani/arch653/winter2010/Freud_Mourningandmelancholia.pdf, accessed 28 July 2018.

85 Lindemann, E., 'Symptomatology and Management of Acute Grief', *The American Journal of Psychiatry* (1944, published online 2006). Available on https://ajp.psychiatryonline.org/doi/abs/10.1176/ajp.101.2.141?journalCode=ajp, accessed 28 July 2018.

86 American Psychiatric Association, *Diagnostic and Statistical Manual of Mental Disorders* (5th edn, Arlington, VA: American Psychiatric Publishing, 2013).

87 Pies, R., 'The Bereavement Exclusion and DSM-4: An Update and Commentary', *Innovations in Clinical Neuroscience*, 11/7–8 (July/Aug 2014), 19–22. Available on: https://www.ncbi.nlm.nih.gov/pmc/articles/PMC4204469/, accessed 28 July 2018.

88 Volpe, A., 'The People Who Can't Stop Grieving', *Independent* (15 November 2016). Available on https://www.independent.co.uk/life-style/health-and-families/people-who-cant-stop-grieving-science-mourning-psychologists-a7416116.html, accessed 28 July 2018.

89 Klass, D., Silverman, P. R., & Nickman, S. L., *Continuing Bonds: New Understandings of Grief* (Washington, DC: Taylor & Francis, 1996).

90 Walter, T., 'How Continuing Bonds Have Been Framed Across Millennia', in D. Klass & E. M. Steffen (eds.), *Continuing Bonds in Bereavement: New Directions for Research and Practice* (London: Routledge, 2018), 43–55.

91 Bonanno, G. A., *The Other Side of Sadness: What the New Science of Bereavement Tells Us about Life after Loss* (New York: Basic Books, 2009), p. 14.

92 Walter (2018), p. 52.

93 Klass & Steffen (2018), p. 4.

94 Bonanno (2009), p. 7.

95 Stroebe, M., & Schut, H., 'The Dual Process Model of Coping with Bereavement: Rationale and Description', *Death Studies*, 23/3 (April/May 1999), 197–224. DOI:10.1080/074811899201046

96 Bonanno (2009), p. 41.

97 Hannon, G., 'Lives Lived – Michael Stanley Kibbee, M.Sc., P.Eng' (1997). Available on: https://cemetery.org/michael-kibbee/

98 According to Marc Saner, the current proprietor of the World Wide Cemetery, serenity and elegance is the USP of this memorial site as opposed to other online memorial spaces that demand payment or feature advertisements.

99 Chayka, K., 'After Death, Don't Mourn, Digitize With Sites Like Eterni.Me', *Newsweek Magazine* (17 August 2014). Available on http://www.newsweek.com/2014/08/29/after-death-dont-mourn-digitize-sites-eternime-264892.html, accessed 28 July 2018.

100 You can view Hong Kong's online memorial garden on https://www.memorial.gov.hk/Default.aspx, accessed 30 July 2018.

101 You can find Legacy.com in the obvious place – Legacy.com.

102 Without having to be a member of Second Life, you can see the memorial on http://secondlife.com/destination/transgender-hate-crime-suicide-memorial, accessed 28 July 2018.

103 Without having to be a member of Second Life, you can see the memorial on http://secondlife.com/destination/peace-valley-pet-cemetary, accessed 28 July 2018.

104 While for reasons of confidentiality I will not link to the page we found, you can locate multiple places online where a guest book or memorial can only be accessed for a fee, such as http://archive.is/t1Lrt

105 Walter, T., 'A New Model Of Grief: Bereavement And Biography', *Mortality*, 1/1 (1996), 7–25. DOI: 10.1080/713685822

106 Originally obtained from http://media.www.redandblack.com/media/storage/paper871/news/2007/11/07/Variety/Facebook.For.The.Great.Beyond-3083145.shtml – link no longer available.

107 Kim, J., 'The Phenomenology of Digital Being', *Human Studies*, 24 (2001), 87–111. Available on: https://www.jstor.org/stable/20011305?seq=1#page_scan_tab_contents, accessed 28 July 2018.

108 Kasket, E., 'Continuing Bonds in the Age of Social Networking', *Bereavement Care*, 31/2 (2012), 62–9. DOI: 10.1080/02682621.2012.710493

109 Roberts, P., 'The Living and the Dead: Community in the Virtual Cemetery', *Omega*, 49/1 (2004), 57–76.

110 Kasket (2012), p. 65.

111 Kasket (2012), 65–6.

112 Office for National Statistics, 'Religion in England and Wales 2011' (11 December 2012). Available on https://www.ons.gov.uk/peoplepopulationandcommunity/culturalidentity/religion/articles/religioninenglandandwales2011/2012-12-11, accessed 30 July 2018. In the 2011 census, Muslims made up the second largest religious group at 4.8 per cent of the population of the UK.

113 Downey, A., 'The U.S. Is Retreating from Religion', *Scientific American* (20 October 2017). Available on https://blogs.scientificamerican.com/observations/the-u-s-is-retreating-from-religion/, accessed 30 July 2018. These data were drawn from the General Society Survey (GSS), which surveys 1,000–2,000 adults a year; note that the US census has not gathered data on religious orientation for some years.

114 Cox, Daniel; Jones, Robert P., 'America's Changing Religious Identity', *2016 American Values Atlas*. Public Religion Research Institute, 9 June 2017. These data were drawn from a 2017 study conducted by the Public Religion Research Institute, which additionally indicated that 69 per cent of Americans considered themselves Christians, 45 per cent saying they attended Protestant churches. Non-Christian religions such as Judaism, Buddhism, Hinduism and Islam collectively make up about 7 per cent of the US population.

115 Gustavsson, A., 'Death, Dying and Bereavement in Norway and Sweden in Recent Times', *Humanities*, 4 (2015), 224–35. DOI: 10.3390/h4020224

116 Walter, T., 'Angels Not Souls: Popular Religion in the Online Mourning for British Celebrity Jade Goody', *Religion*, 41/1 (2011), 29–51.

117 Walter, T., 'The Dead Who Become Angels: Bereavement and Vernacular Religion', *Omega: Journal of Death & Dying*, 73/1 (2016), 3–28.

118 Walter, T., *What Death Means Now: Thinking Critically About Dying And Grieving* (Bristol, UK: Policy Press, 2017).

119 Gustavsson (2015), p. 230.

120 Walter (2016), p. 17.

121 Walter (2017), p. 105.

122 Kasket (2012).

123 Walter (2016), p. 18.

124 Kasket (2012), p. 66.

125 Ibid.

126 Pennington, N., 'You Don't De-friend the Dead: An Analysis of Grief

Communication by College Students through Facebook Profiles', *Death Studies*, 37/7 (2013), 617–35. DOI: 10.1080/07481187.2012.673536

127 Kasket (2012), p. 66.

128 Ibid.

129 Heidegger, M., *Being and Time* (Malden, MA: Blackwell Publishing, 1962).

130 Milian, M., & Sutter, J. D., 'Facebook Revamps Site with "Timeline" and Real-Time Apps', Cnn.com (22 September 2011). Available on https://www.cnn.com/2011/09/22/tech/social-media/facebook-announcement-f8/index.html, accessed 28 July 2018.

131 Pennington, N., 'Tie Strength and Time: Mourning on Social Networking Sites', *Journal of Broadcasting and Electronic Media*, 61/1 (7 March 2017), 11–23. DOI: 10.1080/08838151.2016.1273928

132 Facebook Help Centre (updated 2018). What is a legacy contact and what can they do? [web page]. Available on https://www.facebook.com/help/1568013990080948, accessed 28 July 2018. Note that Facebook's policies and procedures often update and that at time of writing, new procedures are under review.

133 Condliffe, J., 'Will There Ever Be More Dead People Than Living Ones on Facebook?' *Gizmodo* (31 October 2013). Available on https://gizmodo.com/will-there-ever-be-more-dead-people-than-living-ones-on-1455816826, accessed 28 July 2018. This statistic and the following one circulated widely in the popular press, with little information about the methods of calculation used.

134 Cuthbertson, A., 'Dead Facebook Users Will Outnumber the Living by 2098', Newsweek.com (8 March 2016). Available on https://www.newsweek.com/dead-facebook-users-will-outnumber-living-2098-434682, accessed 28 July 2018.

135 Singh, D., 'Facebook's User Growth Stalled in the Second Quarter of 2018', *Business Today* (27 July 2018). Available on https://www.businesstoday.in/technology/news/why-facebook-user-growth-stalled-in-the-second-quarter-of-2018/story/280753.html, accessed 28 July 2018.

136 Numbers drawn from research undertaken at the Oxford Internet Institute's Digital Ethics Lab. See 'Are the Dead Taking Over Facebook: A Big Data Approach to the Future of Death Online. *Big Data and Society*. Available on https://doi.org/10.1177%2F2053951719842540. First published 23 April 2019 and accessed 14 July 2019. See also Ohman, C., & Floridi, L., 'The Political Economy of Death in the Age of Information: A Critical Approach to the Digital Afterlife Industry', *Minds and Machines*, 27/4 (7 September 2017), 639–62. Available on https://link.springer.com/article/10.1007/s11023-017-9445-2, accessed 17 June 2018.

137 See Quiring monuments website on https://www.monuments.com/living-headstones. More will be said about Quiring later in the book.

138 Dr John Troyer is the director for the Centre for Death and Society at Bath University in England and speaks about a range of future possibilities for death and disposition. University of Bath Press Office, 'Researchers Envisage "Future Cemetery"', Phys.org (3 March 2016). Available on https://phys.org/news/2016-03-envisage-future-cemetery.html, accessed 28 July 2018.

139 Phil Wane of Nottingham Trent University demonstrated this technology at the 2018 Death Online Research Symposium, University of Hull, 17 August 2018.

140 This paradoxical nature was originally highlighted to me in an article already cited in this book: Kim, J., 'The Phenomenology of Digital Being', *Human Studies*, 24 (2001), 87–111. Available on: https://www.jstor.org/stable/20011305?seq=1#page_scan_tab_contents, accessed 30 July 2018.

141 Kasket (2012), p. 66.

142 At time of writing, Debra Bassett was a PhD candidate at the University of Warwick, UK, studying digital memories. Website at debrabassett.co.uk, accessed 30 July 2018.

143 I saw a quote from this mother in an unpublished manuscript for a research article authored and provided to me by Debra Bassett – by the time of this book's publication the research is likely to be in the public domain.

144 A good resource for understanding disenfranchised grief is Doka, K. J., *Disenfranchised Grief: New Directions, Challenges, and Strategies for Practice* (1st edn, Champaign, IL: Research Press, 2002). Keep in mind, however, that much has changed – as this chapter illustrates – since the publication of this book.

145 Nicolson, V., 'My Daughter's Death Made Me Do Something Terrible on Facebook', *Guardian* (22 April 2017). Available on https://www.theguardian.com/lifeand-style/2017/apr/22/overcome-grief-daughter-death-died-message-boyfriend-facebook, accessed 28 July 2018. Vanessa Nicolson did not respond to a request for interview made through her agent, hence my paraphrasing the *Guardian* story for this book.

146 Pennington (2013).

147 The names in this anecdote have been changed.

Chapter Three: Terms and Conditions

148 Press Association, 'Hollie Gazzard Attack Was Filmed by Bystanders, Say Police', *Guardian* (19 February 2014). Available on https://www.theguardian.com/uk-news/2014/feb/19/hollie-gazzard-attack-filmed-bystanders-police, accessed 28 July 2018.

149 Mackey, R., 'Taboo on Speaking Ill of the Dead Widely Ignored after Thatcher's Death', *The Lede: Blogging the News with Robert Mackey, New York Times* (10 April 2013). Available on https://thelede.blogs.nytimes.com/2013/04/10/taboo-on-speaking-ill-of-the-dead-widely-ignored-online-after-thatchers-death/, accessed 28 July 2018.

150 Goel, V., 'Facebook Profit Tripled in First Quarter', *New York Times* (23 April 2014). Available on https://www.nytimes.com/2014/04/24/technology/facebook-profit-tripled-in-first-quarter.html, accessed 28 July 2018.

151 Kasket, E., 'Continuing Bonds in the Age of Social Networking', *Bereavement Care*, 31/2 (2012), 62–9. DOI: 10.1080/02682621.2012.710493

152 Price, C., & DiSclafani, A., 'Remembering Our Loved Ones'. [Facebook Newsroom] (21 February 2014). Available on https://newsroom.fb.com/news/2014/02/remembering-our-loved-ones/, accessed 28 July 2018.

153 Pew Research Center, Social media fact sheet (2018). Available on: http://www.pewinternet.org/fact-sheet/social-media/, accessed 28 July 2018.

154 Domonoske, C., 'CDC: Half of All Female Homicide Victims Are Killed by Intimate Partners', *Two-Way: Breaking News from NPR* (21 July 2017). Available on: https://www.npr.org/sections/the-two-way/2017/07/21/538518569/cdc-half-of-all-female-murder-victims-are-killed-by-intimate-partners, accessed 28 July 2018.

155 Office for National Statistics, Compendium: Homicide [web page] (2015). Available on https://www.ons.gov.uk/peoplepopulationandcommunity/crimeandjustice/compendium/focuson-violentcrimeandsexualoffences/yearendingmarch2015/chapter2homicide, accessed 28 Jul 2018.

156 Aslam, S., 'Facebook by the Numbers: Statistics, Demographics & Fun Facts', Omnicore [web page] (1 January 2018). Available on https://www.omnicoreagency.com/facebook-statistics/, accessed 28 July 2018.

157 BBC News, 'Facebook Photos of Hollie Gazzard with Her Killer "Causing Distress"', BBC.co.uk (26 October 2015). Available on http://www.bbc.co.uk/news/uk-england-gloucestershire-34618228, accessed 28 July 2018.

158 I approached John Giacobbi for an interview but received no response, so all material is drawn from the Web Sheriff website.

159 Text from the Web Sheriff website, accessed on https://www.websheriff.com/about-us/ on 28 July 2018.

160 Ibid.

161 This information is drawn from interview with Nick Gazzard.

162 The website for the Centre for the Study of Emotion & Law can be found here: http://csel.org.uk. They primarily focus on vulnerable people and human rights violations.

163 Information about the 2012 Amsterdam Privacy Conference can be found here: http://asca.uva.nl/content/events/conferences/2012/10/2012-amsterdam-privacy-conference.html, accessed 28 July 2012.

164 McDonald, A. M., & Cranor, L. F., 'The Cost of Reading Privacy Policies', *I/S: A Journal of Law and Policy for the Information Society – 2008 Privacy Year in Review Issue* (2008). Available on http://lorrie.cranor.org/pubs/readingPolicyCost-author-Draft.pdf, accessed 28 July 2018.

165 Adams, D., *The Hitchhiker's Guide to the Galaxy* (London: Pan Books, 1979).

166 Balebako, R., Schaub, F., Adjerid, I., Acquisti, A., & Cranor, L. F., 'The Impact of Timing on the Salience Of Smartphone

App Privacy Notices', Security and Privacy in Smartphones and Mobile Devices conference, Denver, Colarado, USA, 12 October 2015. DOI: 10.1145/2808117.2808119. Available on http://www.rebeccahunt.com/academic/timing-balebako.pdf, accessed 28 July 2018.

167 Schaub, F., 'Nobody Reads Privacy Policies – Here's How to Fix That', *The Conversation* (9 October 2017). Available on https://theconversation.com/nobody-reads-privacy-policies-heres-how-to-fix-that-81932, accessed 28 July 2018.

168 'Cambridge Analytica and Facebook: The Scandal and the Fallout So Far', *New York Times* (4 April 2018). Available on https://www.nytimes.com/2018/04/04/us/.../cambridge-analytica-scandal-fallout.html, accessed 28 July 2018.

169 Kavanagh, C., 'Why (Almost) Everything Reported about the Cambridge Analytica Facebook "Hacking" Controversy Is Wrong', *Medium* (25 March 2018). Available on https://medium.com/@CKava/why-almost-everything-reported-about-the-cambridge-analytica-facebook-hacking-controversy-is-db-7f8af2d042, accessed 28 July 2018.

170 Official Journal of the European Union, Regulation (EU) 2016/679 of the European Parliament and of the Council (27 April 2016). Available on http://eur-lex.europa.eu/legal-content/EN/TXT/HTML/?uri=CELEX:32016R0679&from=EN, accessed 28 July 2018.

171 Ibid.

172 This conversation took place on 10 April 2018; this information is drawn from notes taken contemporaneously with conversation.

173 Names in this anecdote have been changed.

174 This is from the 'About Inactive Account Manager' section of the Google website, available on https://support.google.com/accounts/answer/3036546?hl=en, accessed 28 July 2018.

175 More will be said about this in later chapters, but shortly before this book's original publication, Facebook announced a change to their memorialisation features – these changes included increased powers for Legacy Contacts. In a post written by Sheryl Sandberg, the Chief Operating Officer of Facebook, it was explained that 'Legacy contacts can now moderate the posts shared to the new tributes section by changing tagging settings, removing tags and editing who can post and see posts. This helps them manage content that might be hard for friends and family to see if they're not ready.' See 'Making it Easier to Honor a Loved One on Facebook After They Pass Away', published by Sheryl Sandberg on 9 April 2019, available on https://newsroom.fb.com/news/2019/04/updates-to-memorialization/, accessed 14 July 2019.

176 For a useful summary, see Orloff, S., & Frerichs, M. J. (2016, December). Digital Assets After Death: RUFADAA and its Implications. Originally published in Bench & Bar of Minnesota, the official magazine of the Minnesota State Bar Association. Available on https://www.robinskaplan.com/~/media/pdfs/digital%20assets%20after%20death%20rufadaa%20and%20its%20implications.pdf?la=en and accessed on 14 July 2019.

177 For an excellent overview of these Acts and an explanation of how everything developed, see Lopez, A. B. (2016), 'Posthumous Privacy, Decedent Intent, and Post-Mortem Access to Digital Assets'. *George Mason Law Review*. Available on http://www.georgemasonlawreview.org/wp-content/uploads/Lopez_ReadyforJCI.pdf?fbclid=IwAR0C_AyaMKAvjl0oyT_2rgTjO0fxpGyTgvA5i0POsY-9cVAC-Zvv9bkB-h10 and accessed 14 July 2019.

178 Ibid.

179 See the latest update on RUFADAA uptake on the site of the Uniform Law Commission, https://www.uniformlaws.org/committees/community-home?CommunityKey=f7237fc4-74c2-4728-81c6-b39a91e-cdf22, accessed 14 July 2019.

180 John described these questions at a workshop event for User Not Found at the University of Reading on 25 April 2018.

181 At the time of finalising this manuscript, the Law Commission had launched a public consultation on reforming the law of wills on 13 July 2017, a consultation which closed on 10 November 2017 – at time of writing the analysis of responses was ongoing. See https://www.lawcom.gov.uk/project/wills/ for further information.

182 Berlin, J., 'My Appeal to Facebook' [YouTube video] (5 February 2014). Available on https://www.youtube.com/watch?v=vPT28MGhprY, accessed 28 July 2018.

183 Price, C., & DiSclafani, A., 'Remembering Our Loved Ones', Facebook Newsroom (21 February 2014). Available on https://newsroom.fb.com/news/2014/02/remembering-our-loved-ones/, accessed 28 July 2018.

Chapter Four: Behind the Portcullis

184 All names in this story have been changed.

185 English Oxford Living Dictionary definition of 'portcullis', on https://en.oxforddictionaries.com/definition/us/portcullis, accessed 28 July 2018.

186 Morales-Correa, B., 'Wonderful Things: Howard Carter's Discovery of Tutankhamun's Tomb', Ancient History Encyclopedia (4 July 2014). Available on https://www.ancient.eu/article/705/wonderful-things-howard-carters-discovery-of-tutan/, accessed 28 July 2018.

187 United Nations, Universal Declaration of Human Rights, 1948. Available on http://www.un.org/en/universal-declaration-human-rights/, accessed 22 May 2018.

188 Moore, A., Privacy, in International Encyclopedia of Ethics (1) (2013).

189 Hughes, R. L. D., 'Two Concepts of Privacy', Computer Law & Security Review, 31/4 (2015), 527–37. DOI: 10.1016/j.clsr.2015.05.010

190 Huffpost Tech, 'Facebook's Zuckerberg Says Privacy No Longer a "Social Norm"', Huffington Post (18 March 2010). Available on https://www.huffingtonpost.com/2010/01/11/facebooks-zuckerberg-the_n_417969.html?guccounter=1, accessed 29 July 2018.

191 Kirkpatrick, M., 'Facebook's Zuckerberg Says the Age Of Privacy Is Over', ReadWriteWeb (10 January 2010). Available on https://archive.nytimes.com/www.nytimes.com/external/readwriteweb/2010/01/10/10readwriteweb-facebooks-zuckerberg-says-the-age-of-privac-82963.html, retrieved 22 May 2018.

192 Carrascal J. P., Riederer, C., Erramilli, V., Cherubini, M., de Oliveira, R., 'Your Browsing Behavior for a Big Mac: Economics Of Personal Information Online', Proceedings of the 22nd International Conference on World Wide Web (New York: ACM Press, 2013), 189–200.

193 Stieger, S., Burger, C., Bohn, M., & Voracek, M., 'Who Commits Virtual Identity Suicide? Differences in Privacy Concerns, Internet Addiction, and Personality Between Facebook Users and Quitters', Cyberpsychology, Behavior and Social Networking, 16/9 (2013), 629e634. http://dx.doi.org/10.1089/cyber.2012.0323.

194 Kokolakis, S., 'Privacy Attitudes and Privacy Behaviour: A Review of Current Research on the Privacy Paradox Phenomenon', Computers & Security 64 (January 2017), 122–34. Available on http://dx.doi.org/10.1016/j.cose.2015.07.002

195 O'Brien, C., 'Facebook CEO Mark Zuckerberg Begins European Leg of Apology Tour', Los Angeles Times (22 May 2018). Available on http://www.latimes.com/business/technology/la-fi-tn-facebook-zuckerberg-europe-20180522-story.html, accessed 23 May 2018.

196 Fessenden, M., 'How 1960s Mouse Utopias Led to Grim Predictions for Future of Humanity', Smithsonian.com (26 February 2015). Available on https://www.smithsonianmag.com/smart-news/how-mouse-utopias-1960s-led-grim-predictions-humans-180954423/, accessed 29 July 2018.

197 Innes, J. C., Privacy, Intimacy, and Isolation (New York: Oxford University Press, 1992), p. 140.

198 Image © Elaine Kasket 2017, used with permission of Zoe.

199 Hughes, R. L. D., 'Two Concepts Of Privacy', Computer Law & Security Review, 31/4 (August 2015), 527–37. Available on https: doi.org/10.1016/j.clsr.2015.05.010, accessed 23 May 2018.

200 Durham, W. T., & Petronio, S. P., 'Communication Privacy Management Theory: Significance for Interpersonal Communication', in L. A. Baxter, & D. O. Braithwaite (eds.), Engaging Theories In Interpersonal Communication: Multiple Perspectives (2nd edn, Boston, MA: Sage Publications Inc., 2014). DOI: 10.4135/9781483329529.n23

201 Masche, J., 'Explanation of Normative Declines in Parents' Knowledge about Their Adolescent Children', Journal of Adolescence, 33/2 (2010), 271–84. DOI:10.1016/j.adolescence.2009.08.002

202 Thakur, D., 'Sir Arthur Conan Doyle Helped Invent the Curse of the Mummy', Electric Lit [blog] (23 May 2018). Available on https://electricliterature.com/sir-arthur-conan-doyle-helped-invent-the-curse-of-the-mummy-7237bf460528, accessed 29 July 2018.

203 Kaufmann, I. M., & Ruhli, F. J., 'Without Informed Consent?: Ethics and Ancient Mummy Research', *Journal of Medical Ethics*, 36/10 (2010), 608–13. DOI: http://dx.doi.org/10.1136/jme.2010.036608

204 Marchant, J., 'Do Egyptian Mummies Have a Right to Privacy?' *New Scientist* (8 September 2010). Retrieved 22 May 2018 on https://www.newscientist.com/article/mg20727774.600-do-egyptian-mummies-have-a-right-to-privacy/?DC-MP=OTC-rss&nsref=science-in-society, accessed 28 July 2018.

205 Kennedy, C., 'Bad Samaritan or Abuse of a Corpse' [blog post], The Funeral Law Blog (16 November 2016). Available on http://funerallaw.typepad.com/blog/2016/11/bad-samaritan-or-abuse-of-a-corpse.html, accessed 29 July 2018.

206 Jones, I., 'A Grave Offence: Corpse Desecration and the Criminal Law', *Legal Studies*, 37/4 (2017), 599–620. DOI: 10.1111/lest.12163

207 Morris, S., 'Internet Troll Jailed After Mocking Dead Teenagers', *Guardian* (11 September 2011). Available on https://www.theguardian.com/uk/2011/sep/13/internet-troll-jailed-mocking-teenagers, accessed 29 July 2018.

208 Edwards, L., & Harbinja, E., 'Protecting Post-Mortem Privacy: Reconsidering the Privacy Interests of the Deceased in a Digital World', *Cardozo Arts & Entertainment Law Journal*, 32/1 (10 November 2013), 101–47. Available at SSRN: https://ssrn.com/abstract=2267388 or http://dx.doi.org/10.2139/ssrn.2267388

209 Edina Harbinja, 'Post-Mortem Privacy 2.0: Theory, Law, And Technology', *International Review of Law, Computers & Technology*, 31/1 (2017), 26–42. DOI: 10.1080/13600869.2017.1275116

210 Ibid.

211 James Norris provided me with the raw-data results of the 2017 Digital Legacy Survey, which at time of writing had not yet been published on the Digital Legacy Association website.

212 Bleiker, C., 'What Happens to Your Facebook Account after You Die?', DW.com (24 May 2017). Available on http://www.dw.com/en/what-happens-to-your-facebook-account-after-you-die/a-38581943, accessed 29 July 2018.

213 Billingham, P., 'Are Facebook Guilty of Restricting Digital Legacy?' [blog post]. Death Goes Digital [blog] (26 April 2017). Available on http://www.deathgoesdigital.com/blog/digital-legacy-facebook-on-trial, accessed 29 July 2018.

214 AFP/DPA/The Local, 'Update: Parents Can Access Dead Daughter's Facebook, German Court Rules', *The Local DE* (12 July 2018). Available on https://www.thelocal.de/20180712/german-court-to-rule-on-parents-access-to-dead-daughters-facebook, accessed 28 July 2018.

215 Harris, A., 'Digital Legacy Report' [TV broadcast/YouTube video] (8 January 2016). Available on https://www.youtube.com/watch?v=8pi720izNeU, accessed 29 July 2018.

216 This exchange is an exact transcription of the correspondence between Sharon and a Facebook representative, which Sharon emailed to me with permission for use.

217 Full numerical identifier redacted.

218 Personal communication, Jed Brubaker, 7 June 2018.

219 Message thread regarding Facebook notification complaint #5834023 . . ., forwarded with permission for use by Sharon to Elaine Kasket.

220 Message thread regarding Facebook notification complaint #5834023 . . ., forwarded by Sharon with permission for use to Elaine Kasket.

221 Walter (2011).

222 It's been noted that there is a lack of research about religious beliefs and their impact on perceived privacy online, with review of existing literature showing nothing in this regard. The following conference presentation argued that there were theoretical reasons to suspect that cultural and religious beliefs of users may affect the concept of private information, and that this should be further studied. Baazeem, R., 'Social Media Self Disclosure: Individual Digital Privacy and Religion'. Individual paper given at the European Conference on Research Methodology for Business and Management Studies, London, United Kingdom, 2017.

223 Birnhack, M., & Morse, T., 'Digital Remains: Privacy and Memorialisation'. Paper presented at the Information Ethics Roundtable, Copenhagen, Denmark, 2018.

224 Blank, G., Bolsover, G., and Dubois, E. (2014), 'A New Privacy Paradox'. Working Paper, University of Oxford, Global Cyber Security Capacity Centre, 2014.

225 Child, J. T., Petronio, S., Agyeman-Budu, E. A., & Westermann, D. A., 'Blog

Scrubbing: Exploring Triggers That Change Privacy Rules', *Computers in Human Behavior*, 27 (2011), 2017–27.

Chapter Five: Stewarding the Online Dead

226 Indiana Limestone Quarrymen's Association, Bedford, Indiana, *Indiana Limestone: The Aristocrat of Building Materials* [volume 1, sixth edition, June 1920]. Available on the Stone Quarries and Beyond website, http://quarriesandbeyond.org/states/in/pdf/indiana_limestone_the_aristocrat_of_building_materials_1920.pdf, accessed 28 July 2018.

227 The artist for the swan was Don Lawler of Stephensport, Kentucky. His website is donlawlersculpture.com.

228 National Park Service, 'Lincoln Memorial Builders' [blog post] (31 July 2017). Available on https://www.nps.gov/linc/learn/historyculture/lincoln-memorial-design-individuals.htm, accessed 29 July 2018.

229 Image © Elaine Kasket 2018.

230 De Sousa, A. N., 'Death in the City: What Happens When All Our Cemeteries Are Full?', *Guardian* (21 January 2015). Available on https://www.theguardian.com/cities/2015/jan/21/death-in-the-city-what-happens-cemeteries-full-cost-dying, accessed 29 July 2018.

231 Susan's blog is called Breathe, Grieve, Believe and can be found on http://susanfurniss.com/index.html

232 Payne, L., 'The Return of Personal Property to the Bereaved After Disaster: An Exposition of the Key Issues and Guidance for Practitioners'. Paper given at the Eighth International Conference on the Social Context of Death, Dying and Disposal, 2007. Conference programme available on http://www.bath.ac.uk/cdas/documents/ddd8_parallel_sessions.pdf, accessed 29 July 2018.

233 Names in this anecdote have been changed.

234 Bitney Crone, S., 'This Is My Speech That Was Given at Tom's Memorial Service' [blog post] (25 June 2011). Available on http://shanebitneycrone.tumblr.com/post/20131380063/tombridegroommemorialspeech, accessed 30 July 2018.

235 Bloodworth-Thomason, L. (director), *Bridegroom* [documentary film], 2013. Available on https://www.youtube.com/watch?v=RQIIwddt3N4, accessed 30 July 2018.

236 Bejar, A., & Winters, K., 'Introduction to the Compassion Research Team' [PowerPoint presentation] (2011). Available on http://bit.ly/18j2lxP, accessed 30 July 2018.

237 Brubaker, J., 'Death, Identity, & the Social Network', doctoral dissertation, University of California Irvine, 2015. Available on https://escholarship.org/uc/item/6cn0s1xd, accessed 30 July 2018.

238 See https://research.fb.com/programs/fellowship/ for a full list of Facebook Fellows over the years and an explanation of the programme.

239 Information gleaned from Vanessa Callison-Burch's LinkedIn profile and personal communication from Jed Brubaker.

240 Hauser, C., & O'Connor, A., 'Virginia Tech Shooting Leaves 33 Dead', *New York Times* (16 April 2007). Available on https://www.nytimes.com/2007/04/16/us/16cnd-shooting.html?mtrref=www.google.com&gwh=AA75F983D0F6C70E87318D6CF-D486058&gwt=pay, accessed 30 July 2018.

241 Sedghi, A., 'Facebook: 10 Years of Social Networking, in Numbers', *Guardian* (4 February 2014). Available on: https://www.theguardian.com/news/datablog/2014/feb/04/facebook-in-numbers-statistics, accessed 30 July 2018.

242 Ibid.

243 This scene is portrayed in the film *Bridegroom*.

244 Estate Planning Digital Assets, Law Offices of Tyler Q. Dahl, 'Understanding the California Uniform Fiduciary Access to Digital Assets Act' [blog post] Blog for Law Offices of Tyler Q. Dahl (6 August 2017). Available on http://tqdlaw.com/Blog/post/understanding-the-california-uniform-fiduciary-access-to-digital-assets-act, accessed 30 July 2018.

245 These are statistics for the first quarter of 2019, information released from source in June 2019 and obtained from https://www.statista.com/statistics/264810/number-of-monthly-active-facebook-users-worldwide/ on 14 July 2019.

246 See 'Making it Easier to Honor a Loved One on Facebook After They Pass Away', published by Sheryl Sandberg on 9 April 2019, available on https://newsroom.fb.com/news/2019/04/updates-to-memorialization/, accessed 14 July 2019.

247 The website for Quiring Monuments is https://www.monuments.com.

248 Quiring, D., 'Living Headstone ™ – QR Codes Turn Headstones into Interactive Memorials – Quiring Monuments' [YouTube video] (25 May 2011). Available on https://youtu.be/yd2_FG06vnI, accessed 30 July 2018.

249 More information about the 1857 Burial Act can be found at http://www.legislation.gov.uk/ukpga/Vict/20-21/81/contents, accessed 30 July 2018.

250 Photograph in City of London Cemetery © Elaine Kasket 2018.

Chapter Six: The Uncanny Valley

251 Starr, M., 'Eternime Wants You to Live Forever as a Digital Ghost'. CNET Culture (21 April 2017). [Style for media website?] Available on https://www.cnet.com/news/eternime-wants-you-to-live-forever-as-a-digital-ghost/, accessed 27 April 2018.

252 Clute, J., Langform, P., Nicholls, P., & Sleight, G., Encyclopedia of Science Fiction, 3rd edn (London: Gollancz, 2011). Available on: sf-encyclopedia.com, retrieved 26 June 2018.

253 Dick, P. K., The collected stories of Philip K. Dick (New Jersey: Carol Publishing, 1999), pp. xviii-xiv.

254 Bradbury, R., 'Introduction', Analog Science Fact/Fiction 5 (4 April 1973).

255 At the time of writing, this episode is still available on Netflix. Harris, O. (director), & Brooker, C. (writer), 'Be Right Back' [television episode] (2013) from Black Mirror [television series], Channel 4 Television Corporation.

256 Naim, O., The Final Cut [film] (2004), Lions Gate Entertainment.

257 All quotes are direct transcriptions from the film, original script by Omar Naim.

258 Beyond Goodbye media, Beyond Goodbye [film] (2011). Available on https://vimeo.com/38317590, accessed 26 June 2018.

259 King, G., 'Master of Ceremonies' (JBN Memorial), [Vimeo video] (2 September 2012). Available on https://vimeo.com/51035791, accessed 30 July 2018.

260 No author, Stephen Irwin [YouTube video] (5 January 2011). Available on https://www.youtube.com/watch?v=_J6H-6V88ezc, accessed 26 June 2018.

261 Rotten Tomatoes' website can be found at https://www.rottentomatoes.com

262 See https://www.rottentomatoes.com/m/final_cut for all reviews. Some still link to the original full-text reviews.

263 2004 (US) domestic grosses can be viewed on https://www.boxofficemojo.com/yearly/chart/?yr=2004&p=.htm, accessed 30 July 2018.

264 Grosses for Robin Williams films can be viewed on https://www.boxofficemojo.com/people/chart/?view=Actor&id=robinwilliams.htm, accessed 30 July 2018.

265 All quotes are transcribed directly from the episode, the original screenplay by Charlie Brooker.

266 Business Insider Intelligence, '80% of Businesses Want Chatbots by 2020', Businessinsider.com (14 December 2016). Available on http://uk.businessinsider.com/80-of-businesses-want-chatbots-by-2020-2016-12, accessed 30 July 2018.

267 Kihlstrom, G., 'When It's Time to Consider Chatbots as Part of Your Customer Experience Strategy' [blog post], Forbes.com. (1 June 2018). Available on https://www.forbes.com/sites/forbesagencycouncil/2018/06/01/when-its-time-to-consider-chatbots-as-part-of-your-customer-experience-strategy/#483421b44d8a, accessed 30 July 2018.

268 For a fascinating and comprehensive story about Eugenia Kuyda and the Roman Mazurenko chatbot: Newton, C., 'Speak, Memory', TheVerge.com. (n.d.). Available on https://www.theverge.com/a/luka-artificial-intelligence-memorial-roman-mazurenko-bot#conversation1, accessed 30 July 2018.

269 Savin-Baden, M., & Burden, D., 'Digital Immortality and Virtual Humans', paper presented at Death Online Research Symposium, University of Hull 15–17 August 2018. At time of writing a fuller version of the paper was due to be published in Postdigital Science and Education in the latter part of 2018.

270 Mathur, M. B., & Reichling, D. B., 'Navigating a Social World with Robot Partners: A Quantitative Cartography of the Uncanny Valley', Cognition, 146 (2016), 22–32.

271 I first encountered the concept of the 'uncanny valley' myself when attending a presentation by Debra Bassett. Basset, D., 'Who Wants to Live Forever? Living, Dying and Grieving in Our Digital Society'. Individual paper presented at CyPsy22, University of Wolverhampton, Wolverhampton, England, 27 June 2017.

272 Plunkett, John, 'Black Mirror Nets Nearly 1.6m Viewers', *Guardian* (12 February 2013). Available on https://www.theguardian.com/media/2013/feb/12/black-mirror-charlie-brooker-tv-ratings, accessed 30 July 2018.

273 Bramesco, C., 'Every *Black Mirror* Episode, Ranked from Worst to Best', *New York Magazine: Vulture* (29 December 2017). Available on http://www.vulture.com/2016/10/every-black-mirror-episode-from-worst-to-best.html, accessed 30 July 2018.

274 Tate, G., 'Charlie Brooker and Hayley Atwell Discuss "Black Mirror"', *Time Out London* (31 January 2013). Available on https://www.timeout.com/london/tv-and-radio-guide/charlie-brooker-and-hayley-atwell-discuss-black-mirror, accessed 30 July 2018.

275 Ibid.

276 'The Future Starts Here' was an exhibition at the V&A museum in London that ran through November 2018. For information on current and past exhibits, see vam.ac.uk

277 See http://eterni.me

278 At time of writing, Paula Kiel's research webpage could be found at https://onlineafterdeath.weebly.com, accessed 30 July 2018.

279 Shavit, V., 'My Retirement Post' [blog post], *Digital Dust* (2 March 2018). Available on http://digital-era-death-eng.blogspot.com/2018/03/my-retirement-post.html, accessed 30 July 2018.

280 Carroll, E., & Romano, J., *Your Digital Afterlife* (San Francisco, CA: Peachpit Press/New Riders, 2010).

281 For the list, see http://www.thedigitalbeyond.com/online-services-list/

282 Because some websites provide more than one service, the percentages don't add up to 100.

283 Nicolson, V., 'My Daughter's Death Made Me Do Something Terrible on Facebook', *Guardian* (22 April 2017). Available on https://www.theguardian.com/lifeandstyle/2017/apr/22/overcome-grief-daughter-death-died-message-boyfriend-facebook, accessed 28 July 2018.

284 Brubaker, J., 'Death, Identity, & the Social Network', doctoral dissertation, University of California Irvine, 2015. Available on https://escholarship.org/uc/item/6cn0s1xd, accessed 30 July 2018.

285 Before its closure, Perpetu was featured in a 2013 TechCrunch article: Shu, C.,

'Perpetu Lets You Decide What Happens to Your Online Accounts After You Die', TechCrunch.com (5 December 2013). Available on https://techcrunch.com/2013/12/04/perpetu/, accessed 30 July 2018.

286 In 2011, the CDAS conference at the University of Bath focused on death and digital technologies.

287 At the time of writing, Dr Stacey Pitsillides is an academic at the University of Greenwich. For more on her Digital Death work, see http://www.digitaldeath.eu

288 For more information about Debra's work, see her website at http://debrabassett.co.uk

289 Sunday Sequence, 'Digital Afterlife', BBC Radio Ulster (4 February 2018). Available on https://www.bbc.co.uk/programmes/p05x1wrs, accessed 30 July 2018.

290 To view the video of Goldberg interviewing BINA48, see https://www.youtube.com/watch?v=j1vB7OHe4EA

291 Stacey Pitsillides has organised Love after Death events in various locations; it is part of her ongoing efforts to engage the public in discussions of death and dying, including considerations of the digital age. The event in which I participated took place at Redbridge Library in May 2018. More information can be found on http://www.digitaldeath.eu/updates/, accessed 30 July 2018.

292 Paula did not name this developer.

293 This film was based on a novel by Cecelia Ahern. LaGravenese, R. (director, screenwriter), & Rogers, S. (screenwriter), *P.S. I Love You* [film], Warner Bros (2007).

294 The SafeBeyond website can be found on https://www.safebeyond.com. SafeBeyond did not respond to a request for interview in 2018, and in June 2019 it became clear why this might have been. After a period of inactivity, a post appeared on Facebook and Twitter on 11 June 2019, reading: 'Hello SafeBeyond members! We wanted you to be the first to know that SafeBeyond has been acquired, and will be continuing to offer members the ability to share both important and special parts of their lives with their loved ones. We are feverishly working through this transition and wanted to let you know that current members will not lose their prior messages and will still be able to record new messages and save important information. If you have any questions or concerns, feel free to reach out to us at info@safebeyond.

com. Thank you for your patience!' See 11
June 2019 post on https://www.facebook.
com/safebeyond/ accessed 14 July 2019.
This must have come to a relief to users
who received a worrying message just a
few days before encouraging them to
download their information: see https://
twitter.com/djbassett2/
status/1137718201872191489 for a copy of
that message.

295 Griffin, A., 'Facebook Friend Requests
from Dead People Hint at Horrifying
Truth of "Profile Cloning"', *Independent*
(10 April 2017). Available on https://www.
independent.co.uk/life-style/gadgets-and-
tech/features/facebook-friend-request-
dead-people-cloning-hack-hoax-safe-
problem-a7676361.html, accessed 30 July
2018.

296 Crown Prosecution Service, 'Stalking and
Harassment' (23 May 2018). Available on
https://www.cps.gov.uk/legal-guidance/
stalking-and-harassment, accessed 30 July
2018.

297 My Farewell Note had its soft launch in the
latter part of 2017. See the website at
https://www.myfarewellnote.com

298 *Rutland & Stamford Mercury* staff writer,
'App Allows People to Say Bye Before
Death', *Rutland & Stamford Mercury* (11
November 2017). Available on https://
www.stamfordmercury.co.uk/news/
app-allows-people-to-say-bye-before-
death-1-8238334/, accessed 30 July 2018.

299 Barrett, P., 'My Old Dog Singing to
Eastenders' [YouTube video] (8 June 2016).
Available on https://www.youtube.com/
watch?v=Vs9M1dSTX0w, accessed 30 July
2018. The bit where the voice of Peter's
father can be heard is at approximately 1
minute 25 seconds.

300 Marius's profile for MIT Bootcamp can be
found at https://bootcamp.mit.edu/
entrepreneurship/speakers/
mariusursache/, accessed 30 July 2018.

301 The Founders Forum website is at https://
ff.co/#/, accessed 30 July 2018.

Chapter Seven: The Voice of the Dead

302 LeBrecht, N., 'Exclusive: My Life With
Maria Callas', Slippedisc.com (31 May
2016). Available on http://slippedisc.
com/2016/05/eclusive-my-life-with-
maria-callas/, accessed 30 July 2018. This
post presented an excerpt from a recently
published memoir by Jacques Leiser.

303 Goss, N., & Hoffman, E., *Tearing the*
World Apart: Bob Dylan and the
Twenty-First Century (Jackson, MI:
University Press of Mississippi, 2017), p.
33.

304 Neary, L., 'Maria Callas: Voice of Perfect
Imperfection', NPR.org. (15 February
2010). Available on https://www.npr.org/
templates/story/story.php?story-
Id=123612228, accessed 30 July 2010.

305 At time of writing, this material was on
BASE Hologram's website, available on
https://basehologram.com/productions,
accessed 30 July 2018.

306 McDonald, S. N., 'Are Holograms a Creepy
Way to Honor Fallen Icons Like Michael
Jackson?', *Washington Post* (19 May 2014),
retrieved 9 August 2018. Available on:
https://www.washingtonpost.com/news/
morning-mix/wp/2014/05/19/are-holo-
grams-a-creepy-way-to-honor-fallen-
icons-like-michael-jackson/?noredirect
=on&utm_term=.2df01b90e4aa, accessed
30 July 2018.

307 Bukszpan, D., 'Life After Death: Musicians
Are Coming Back to the Stage Thanks to
Holographic Technology', CNBC.com (12
August 2017). Available on https://www.
cnbc.com/2017/08/11/musicians-are-com-
ing-back-to-life-thanks-to-holograms.
html, accessed 9 August 2018.

308 Ibid., accessed 30 July 2018.

309 Reported on Talking Metal and quote
reproduced on http://ultimateclassicrock.
com/dio-hologram-creator/

310 For full article, inclusive of Alex Orbison
quote, see Daly, R., 'We Got a Sneak
Preview of Roy Orbison's Hologram Tour
and It's Pretty Mind-Blowing', NME Blogs
(1 February 2018). Available on https://
www.nme.com/blogs/nme-blogs/
roy-orbison-hologram-heritage-
tours-2233785, accessed 30 July 2018.

311 See https://diana-award.org.uk/award/
about/ for more information about the
Diana Award.

312 Lucy's blog can be found at http://www.
lucy-watts.co.uk, accessed 30 July 2018.

313 The BBC3 documentary *Rest in Pixels* was
uploaded to the site on 11 May 2016. It is
no longer available to view on BBC iPlayer
but can be seen on YouTube, available on
https://www.youtube.com/watch?v=W1K-
swKZxtaA, accessed 30 July 2018.

314 The question set and associated data from
the 2017 Digital Death Survey, results not
yet published, were provided to me
directly by James Norris.

315 The Bob Monkhouse prostate cancer ad can be viewed on YouTube, available on https://www.youtube.com/watch?v=qux-OMX1jl1w, accessed 30 July 2018.

316 BBC News staff, 'Monkhouse to Appear in Cancer Ads', BBC News.co.uk. (12 June 2007). Available on http://news.bbc.co.uk/1/hi/health/6743261.stm, accessed 30 July 2018.

317 The current Digital Legacy Association team can be seen on https://digitallegacy-association.org/about/digital-legacy-asso-ciation-board/

318 Personal communication from James Norris and update provided on the deadsocial.org website, accessed 30 July 2018. The MyWishes website is due to be launched in late summer/early autumn 2019 and is on https://mywishes.co.uk, accessed 14 July 2019.

319 Screenshot of Jason Noble's LinkedIn profile obtained from Greg King. Permission to use content obtained from Jason Noble's widow and executrix.

320 King, G., 'Remembering Jason Noble: Superhero', LEO Weekly (15 August 2012). Available on https://www.leoweekly.com/2012/08/remembering-jason-noble-superhero/, accessed 30 July 2018.

321 Rolling Stone staff writer, 'Rodan Singer Jason Noble Dead at 40', Rolling Stone (6 August 2012). Available on https://www.rollingstone.com/music/music-news/rodan-singer-jason-noble-dead-at-40-107083/, accessed 30 July 2018.

322 The CaringBridge.org page for Jason is no longer viewable by the public.

323 Ashby, H. (director) & Higgins, C. (writer), Harold and Maude [film], Paramount Pictures (1971).

324 DeGroot, J. M., 'For Whom the Bell Tolls: Emotional Rubbernecking in Facebook Memorial Groups', Death Studies, 38/1–5 (2014), 79–84.

325 Gach, K., Fiesler, C., & Brubaker, J., "'Control Your Emotions, Potter": An Analysis of Grief Policing on Facebook in Response to Celebrity Death', PACM on Human Computer Interaction, 1/2 (2017), article 47. DOI: 10.1145/3134682

326 Berkowitz, P., 'Carry Him Forward: Remembering the Life, Art and Inspiration of Jason Noble', LEO Weekly (15 August 2012). Available on https://www.leoweekly.com/2012/08/carry-him-forward-2/, accessed 30 January 2018.

327 Remembering Jason Noble [Facebook group]. Available on https://www.facebook.com/groups/269736773137765/, accessed 30 July 2018.

328 Email from Gregory King to Elaine Kasket, 14 July 2018.

329 Harvey, R., 'The Unquiet Grave', Journal of the English Folk Dance and Song Society, 4/2 (December 1941), 49–66. Available on https://www.jstor.org/stable/4521181

330 The video for 'Lazarus' can be seen on https://www.youtube.com/watch?v=y-JqH1M4Ya8

331 Bowie, D. 'Lazarus', from the album, Blackstar, produced by David Bowie and Tony Visconti. Columbia Records (2016).

332 Ibid.

Chapter Eight: Remembering Zoe

333 Information Commissioner's Office, 'What Rights Do Children Have?', ICO.org.uk. (2018). Available on https://ico.org.uk/for-organisations/guide-to-the-general-data-protection-regulation-gdpr/children-and-the-gdpr/what-rights-do-children-have/, accessed 30 July 2018.

334 Steinberg, S., 'Sharenting: Children's Privacy in the Age Of Social Media', 66 Emory L.J. 839 (2017). Available on https://scholarship.law.ufl.edu/cgi/viewcontent.cgi?article=1796&context=-facultypub, accessed 30 July 2018.

335 Ibid., p. 1.

336 Cherney, M. A., 'Facebook Stock Drops Roughly 20%, Loses $120 Billion in Value after Warning That Revenue Growth Will Take a Hit', Marketwatch (26 July 2018). Available on https://www.marketwatch.com/story/facebook-stock-crushed-after-revenue-user-growth-miss-2018-07-25, accessed 28 July 2018.

337 Bowie, D. 'Lazarus', from the album, Blackstar, produced by David Bowie and Tony Visconti. Columbia Records (2016).

338 Sarie Miell is a designer based in East London – her website is fusedandfired.com.

339 Image © Elaine Kasket 2018. Zoe and her grandmother at the swan memorial to Elizabeth, James and their daughter Mary Ann. Their images are reproduced here with the permission of Zoe and of Beth Rudwell.

Picture Credits

Index

Note: page numbers in **bold** refer to diagrams.

Abrams, Rachel 26
access to digital remains 28–9, 63–7, 80,
    90–2, 94–7, 102–8, 119–28, 131–2,
    156, 197
Adams, Douglas 84
advanced-care directives 209–10
afterlife 2, 54–5, 126, 215
    digital 199
Age of the Friend 65
agency, loss of 67
agnosticism 55
Alexa 189–90
always-ons 16–17, 18, 226
Amazon 94, 158, 219
    deceased account closure 87–8
Amazon Echo 189–90
ancestor worship 41, 56
angels 57
    dead as 55–6, 126
    social agency of 56
antibiotics 9
anxiety, death 8–9, 243–4
AOL search data 31
Apple 219
Apple Music 52
Arnold, Thelma 32
Ashby, Hal 223
ashes 160
assets 18–24, 93
    tangibility 19, 91–2, 115–16
    value 19
    see also digital assets
atheism 54, 55
augmented reality technologies 61
Austen, Jane 9
autobiographies, digital **18**, 24–5, 53,
    58–9, 199, 243–4
Autographer 17

autonomy 119
avatars 167, 169, 175–9, 181, 199, 200–1

'backing up' 227
Bakewell, Joan 19
Barrett, Peter 188–9, 192, 194–7, 214
barrier between life and death, traversing
    the 57–8
BASE Entertainment 204, 207
Bassett, Debra 181, 182
BBC 38, 77, 218
BBC3 214
BBC Points West 79
BBC Radio 189, 212
BBC TV 121, 124
Being 58
Being-in-the-world 58
Berlin, Jesse 99–100
Berlin, John 99–100
Beyond Goodbye (2011) 169–70
Billboard Music Awards 2014 205, 218
Billingham, Pete 120
BINA48 (robot) 182, 192, 202
biographies, unauthorised digital **18**,
    25–7, 71
Bitney Crone, Shane 149, 150–1, 155–6
Black Mirror, 'Be Right Back' (2013) 169,
    173–8, 180–2, 186, 202
blog scrubbing 128
blogs (weblogs) 13, 140, 212–13
    and curators 15
    parental 235
    posthumous 219–20
bodily privacy 109, 117
Bonanno, George 43
Boston 10
boundaries
    public–private 110

setting 111, 113, 120
shifting 129, 132
turbulence 113
Bowie, David 224, 231–4, 236–8
Bradbury, Ray 168, 172, 177
brain-to-digital mind uploading 182
Brannen, Kate 32
Branson, Richard 196
Bridegroom, Tom 149, 150–1, 155–6
Brooker, Charlie 177, 178
Brookwood Cemetery 10
browsing history, value 110
Brubaker, Jed 151–9, 162, 166, 180, 187, 200, 224, 226
bullying, cyberbullying 37–8, 157
Burial Act 1857 165
burials, costs of 160
Burks, Gary 163–5, 227
businesses, online, shutting down after death 23–4

Callas, Maria 203–4, 206, 207
Callison-Burch, Vanessa 153, 155, 157
Cambridge Analytica 85, 99, 111, 151, 200, 225
Campos, Juliana 38
cancer 139–44, 184, 217–18, 221–3, 228, 230, 233
'care culture' of mourning 42, 44–5
caretakers of the dead 136–9
CaringBridge 222–4
Carnarvon, Earl of 108, 114, 132
Carroll, Evan 179–80
Carter, Howard 108
Carter, Mrs L. 5
Catholicism 54, 215
CDAS (Centre for Death and Society), Bath University 22, 60–1, 107, 181, 227
CDC (Centers for Disease Control and Prevention) 76
celebrity deaths 224, 231, 233
cemeteries 10, 11, 156, 159–66
and digital memorials 160–1
enhanced/hybrid 60–1, 161–2
leasehold graves of 165–6
and memorialisation 159–60
permanent graves (perpetuity) of 164–6
Victorian 164–5
see also virtual cemeteries
Centers for Disease Control and Prevention (CDC) 76
Centre for Death and Society (CDAS), Bath University 22, 60–1, 107, 181, 227

Centre for the Study of Emotion & Law 81
CGI (computer-generated imagery) 205
chatbots 175, 181–2
children
and consent issues 235
digital footprints 238
privacy rights 234–9
'protecting' from death 36
Chilon of Sparta 71
China 41, 56
choice 239
and children 112, 113, 234, 238
and continuing bonds 44
and cremation 160
and deceased Facebook accounts 100
and end-of-life planning 211, 212, 214
Christians 9, 54, 149–51, 215
fundamentalist 149–50
Church 165
circle of trust 129
City of London Cemetery & Crematorium 163–6, **166**, 216, 227
closedness 16
collectivist cultures 90
commercialisation of death 48
communications privacy management (CPM) theory 113, 128
communities
of grief 141, 142, 144, 149
of mourners 65
Complex Ehlers-Danlos Syndrome 208
computer-generated imagery (CGI) 205
Conan Doyle, Arthur 114
continuing bonds theory 3–5, 41–2, 44, 58–9, 61
digital disruption of 63–9
control issues 104
and children 238–9
and digital remains 63, 65–9, 72–81, 164
and end-of-life planning 211, 212, 214
and families 73–81, 88, 94, 96
and gravestones 163–4
and the Information Age 14
and next of kin 155–7
coping paradoxes 38
copyright laws 92–3, 94, 118
Coutinho, Dolores Pereira 38
CPM (communications privacy management) theory 113, 128
Cranor, L. F. 83
cremation 160
criminal activity, and the invisibility of the Internet 86

Cummings, Rebecca 29
curators 15, 18, 24, 247–8
cyberbullying 37–8, 157

data *see* hyper-personal data; personal data
data protection 84–6, 88–9, 93, 131, 234
data storage crises 166
de Montaigne, Michel 236
dead bodies 11, 116
Dead Social 188, 215, 218–19
dead, the
    caretakers 136–9
    need to talk to 2, 5–7, 51–4, 56–8
    possessions of xi–xix, **xii**, 48
    respect for 114
    sentient 54, 55
    as stalkers 194
    staying connected to the 3–7
    stewarding 133–66
death
    celebrity 224, 231, 233
    difficult 249
    Irish approaches to 36–7
    premature 49–54, 61–3, 70–81, 99–100,
        102–8, 119–29, 139–49, 157, 183–4,
        189–90, 199, 221–3, 249
    'protecting' children from 36
    talking about 245
    see also life and death; 'second death'
death anxiety 8–9, 243–4
Death Goes Digital podcast/blog 120
'death – the human experience'
        exhibition, Bristol Museum & Art
        Gallery, 2016 8
death notifications 67
death phobia, Western 8
death taboos 8–9, 197
'death tech space' 188, 200, 202, 215, 218
death-denial mode 8–9
depression 40, 191
*Diagnostic and Statistical Manual of
    Mental Disorders* 40
Dick, Philip K. 168, 172, 175, 178
digital afterlife 199
digital archives **18**, 27–9
    stewardship 145–8
digital assets 18–24, **18**, 78–9
    and digital estate planning services
        179–80
    and the Digital Legacy Association 219
    rights to 90, 91–2, 96–9
Digital Assets Inventory Worksheet 20–1
digital autobiographies **18**, 24–5, 53, 58–9,
    199, 243–4

Digital Beyond (website) 180–1
digital biographies, unauthorised **18**,
    25–7, 71
digital death services 179–81, 183, 188,
    195
    see also specific services
digital disruption, of continuing bonds
        theory 63–9
'digital divide' 13
digital dossiers **18**, 29–32
Digital Dust (blog) 179
digital executors 246
digital footprints 7, 29, 32–3, 52–3
    of always-ons 17
    children's 238
    of curators 15
    and death notifications 67
    of digital pragmatists 15
    of hermits 14
    and posthumous messages 199
    sparse 220
    unpleasant 68
digital immigrants xx–xxi, xxii, 2, 48, 189
    and curators 15
    and digital assets 19
    and hybrid cemeteries 161
    and land law reform 98
    and online grief 34
digital immortality 166–7, 169, 178–9,
    181, 199–201
digital legacies 19–20, 102, 105, 121, 144,
    179, 188, 207, 214, 219, 234, 245
    access to 66–7, 125, 148
    of Allem Halkic 37
    and digital dossiers 32
    and digital footprints 29, 32
    and Facebook's Legacy Contact 96, 97
    guide to 243–51
    of Hollie Gazzard 69, 71, 77–81, 90,
        99
    legacy websites 170
    parental access to 102–8
    and photography 145
    planning 183, 219, 244
    as product of living actions 200
    substantial 48, 52
    and terms and conditions 99
    and unauthorised biographies 26
    unpleasant 67–8
    and the World Wide Cemetery 46
    see also digital remains
Digital Legacy Association 19, 22, 119,
    179, 188, 216, 219
digital memorials 160–1

digital minds, brain-to-digital mind
  uploading 182
digital natives xx, xxi–xxii
  and always-ons 16, 17
  and Facebook 51, 53
  and hybrid cemeteries 161
  and the loss of digital remains 64
  and online grief 35
  and post-humous messaging systems
    199
'digital pragmatists' 14–15
digital reflections 44, 64–5, 201, 244, 248
digital remains 242–51
  access to 28–9, 63–7, 80, 90–2, 94–7,
    102–8, 119–28, 131–2, 156, 197
  content of 164
  control issues 63, 65–9, 72–81, 164
  and the dead's right to privacy 116–27
  fear of losing 64
  guide to 243–51
  hyper-personal nature 115, 118, 126,
    130, 234
  and the living's right to privacy
    127–32
  permanence 165–6
  stewardship of 138–66
  unpleasant 67–8
  see also digital legacies
digital self 64
digital wills 89–90, 96–7, 246
digital-age citizens
  typology 14–18
  see also always-ons; curators; digital
    pragmatists; hermits; life loggers
digitally stored information, fragility of
  64, 226–7, 236, 249
dignitary privacy 112–13
DiNucci, Darcy 12
Dio, Ronnie 206
disaster victims, personal property 48
disclosure of information 128–9
disenfranchised mourners 65
Dublin 34
Duffy, Sean 118
Dylan, Bob 203
dystopia 17

Edison, Thomas 5
Eggers, Dave, The Circle 17
Egyptian mummies, right to privacy
  114–15
Electronic Communications Privacy Act
  (1986) 28
Ellsworth, John 28–9, 90–2, 127–8

Ellsworth, Justin 27–9, 90–2, 127–8, 130
Elysium, new 7
emails 225
  stewarding digital archives of the
    deceased 145–6, 147
EMI 203
Encyclopedia of Science Fiction, The 168
end-of-life planning 209–14
Engelberts, Lernert 31
Eternime app 166–7, 178–9, 181, 196–202
European Union (EU) 85, 88–9, 117, 131,
  234
executors, digital 246
Eyellusion 206

Facebook 94, 138, 147, 200, 213, 218–20
  and access issues 64, 65, 102–7, 120–6,
    122, 124
  birth 13
  and celebrity deaths 224, 231, 233
  Christian sites 149–50
  closed group support pages 141–4
  and commercial incentives of the
    cemetery function 159
  Compassion Research Team 151–9, 187
  and control issues 65–9, 73–5, 155–7
  and the data protection scandal 85, 111
  and deactivated accounts 140–1, 144
  and death notifications 67
  and the deletion of content 64, 149–51,
    156
  and the demise of privacy 109–10
  and digital autobiographies 24–5, 51,
    53, 58–9
  and digital footprints 51, 53
  and Facebook Fellowship grants 152
  and help forums 185–6
  Irish uptake 34
  and Legacy Contacts 25, 59, 89, 95–8,
    119, 153, 155–7, 246
  and logging into deceased person's sites
    37–8
  and 'Look Back' videos 99–100
  and memorial groups 224–6
  and memorialisation of profiles 25, 38,
    49–54, 56–7, 59–60, 65–7, 75–80,
    99–100, 102–7, 123–4, 153–4, 159,
    165–6
  and Messenger 121, 147
  and mourning 2
  and next of kin's access to 122, 124
  and ownership rights 92
  and parental access to deceased
    accounts 102–7, 120–6

and personal data sharing 110–11
and photography 76–8, 80, 92–4,
    103–4, 106
potential fragility of 227, 236
and privacy settings 92–3
and the Reconnect feature 180
and scandals 60, 111, 151, 225
and sharenting 235
and surveillance technology 30
talking to the dead on 51–4, 56–8
and terms and conditions 85, 86
and Timelines 25, 58, 95–6, 127, 157
and traditional cemeteries 161
and trolling 71–3, 116
and upsetting content 76–8, 80, 92,
    103–4, 106
user numbers 59, 60
and the User Operations department
    123
and the Wall 50
Fallujah 28
families
    and access to digital remains 65–7
    and control issues 65–7, 73–81, 88, 94,
        96
    and the deletion of Facebook profiles
        149–51, 156
    and grief 35–6
    see also next of kin; parents
fathers, absent 68–9
Federal Bureau of Investigation (FBI) 72
Federal Court of Justice 120
Final Cut, The (2004) 169–73, 176, 177–8,
    181
Founders Forum 196
Fox, Kate 4–5, 7
Fox, Margaret 4–5, 7
Freud, Sigmund 3–4, 41
    'Mourning and Melancholia' (1917)
        3–4, 11, 39
friends
    Age of the Friend 65
    and online access to the deceased 65,
        102–3
Friends of Jonnie Sue group 141–4
Friendster 49, 140
funerals 242
    playlists 215–16

Gazzard, Chloe 71–2, 73–4, 78, 84, 88
Gazzard, Hollie 24–5, 69–81, 83–6, 89–90,
    92, 99–100, 103–4, 121, 123, 157
Gazzard, Nick 24–5, 71–80, 84, 86, 88,
    99–100, 104

GCHQ see Government Communications
    Headquarters
General Data Protection Regulation
    (GDPR) 2018 85, 89, 117, 234
Generation AO 16
Germany 120–1, 137
Giacobbi, John 79–80
Glasgow Association of Spiritualists 3
Gmail 105
God 55, 150
Goldberg, Whoopi 182, 192
GoneNotGone 192–5, 197, 201, 214, 219
Google 26
    Inactive Account Manager (IAM)
        feature 89, 94–5, 119, 246
    open-source software 175
    surveillance technology 30
Google Drive 158
Google Glass 17
Government Communications
    Headquarters (GCHQ) 72
graves 156
    expiry 137
    grave-recycling practices 137
    replacement by online sites 143
gravestones 133–8, **136**, 159–60,
    163–4
    construction materials 159–60
    inscription control 163–4
    'Living Headstones' 161–2, 164
Graziano, Michael 182
grief 249
    anatomy of online grief 34–69
    communities of 141, 142, 144, 149
    complicated 35, 39, 69
    dual process model of 43
    medicalisation 40
    oscillations of 43–4
    pathologisation 4
    stages of 39–41, 43–4
    styles of 43
    support with 141–4
    theories of 3–5
    work of 4, 41
    see also mourning
group privacy 127–32
Guardian (newspaper) 66
Guns N' Roses 215–16

hackers 22–3, 186
Halkic, Ali 37–8, 56
Halkic, Allem 37–8, 56
Halkic, Dina 38
Hanks, Tom 17

Harbinja, Dr Edina 89–94, 96–8, 108, 115, 118–20, 131, 151–2
hard copy, storing digital content as 91, 236–7, 249–50
Harold and Maude (1971) 223
Heidegger, Martin 58
hermits 14, 32, 242
Herrmann, Ulrich 120
Highgate Cemetery 10
HIStory tour 1997 218
hoaxers 7
Holiday, Billie 205, 206
Hollie Gazzard Trust 73
Hologram USA 204–5
holographic projections, musicians 204–8, 218
Home Office 165
homosexuality 149, 150–1, 155, 156
Hong Kong 46
Houston, Whitney 205, 206
Huffington, Arianna 196
hyper-personal data 115, 118, 126, 130, 234

I Love Alaska (2009) 31–2, 81–2, 110
immortalists 8
immortality 8
  digital 166–7, 169, 178–9, 181, 199–201
  fantasy of 8
  forgetting/leaving behind 250–1
Independent (newspaper) 185–6
individualistic cultures 74, 90
Industrial Revolution 9, 11
industrialisation 9–10
infant mortality 9
information
  disclosure of 128–9
  personally identifiable 117, 127
  private 110
  public 110
  sensitive identifiable 117, 127
Information Age, control of the 14
Information Ethics Roundtable 126
information-sharing 128–9
  see also sharenting
informational privacy 109, 116–17, 127
Instagram 52, 77, 235
intellectual property 92, 93–4, 118, 206
interconnectivity 58
Internet 13, 152, 213, 215
  invisibility of the 86
  and the proliferation of information 81
intestate persons 28, 90–1, 93, 119
Iraq 28, 29

Ireland 34–7
Israel 145

Jackson, Michael 205, 206, 218
J's Hospice 209–11, 212

Kensal Green Cemetery 10
Kibbee, Mike 45
Kiel, Paula 179, 183–8, 192, 197
King, Greg 169, 220–2, 224–30, 236
Kingdom 150
Kübler-Ross, Elisabeth 39, 40–1, 43–4
Kuyda, Eugenia 175–6

laptops 93–4, 106, 146–7, 227
law 18–20, 81–2, 88–93, 96–9, 155
  black-letter 91, 115
  copyright 88, 92–3, 94, 118
  data protection 88–9, 234
  of the land (local law) 88, 92
  and post-mortem privacy 131
  and privacy 89, 109, 131
  of succession 19, 89, 90, 93, 118
Law Commission 98, 131
legacies 220, 230, 241
  see also digital legacies
Legacy.com 46
'legal person', extension beyond death 118
Legge, Walter 203
letter-writing xiii–xix, 119
'letting go' 4
liability 86
life and death
  separation of 11
  traversing the barrier between 57–8
life expectancy 9
life loggers 17, 18, 24
lifehacker (blog) 30
Lincoln Memorial, Washington, DC 135, 136
Lindemann, Erich 39–40
LinkedIn 199, 220–1, 223, 225, 229
'Living Headstones' 161–2, 164
London 10
London Cemetery Company 10, 165
longevity 9
longevity scientists 8
loss 139–40, 189
Love After Death event 2018 207
Lucy's Light 212, 213

MacBryde, Natasha 116, 118
McDonald, A. M. 83
Maslin, Asher 70–1, 77, 80, 92

Mazurenko, Roman 175–6
McIlroy, Steve 188–92, 194, 196, 214
mediums 5, 56
memorial groups, Facebook 224–6
memorial websites 140
memorialisation 159
    of Facebook profiles 25, 38, 49–54,
        56–7, 59–60, 65–7, 75–80, 99–100,
        102–7, 123–4, 153–4, 159, 165–6
memorials
    digital 160–1
    unrealistic portraits 68–9
'memory culture' of mourning 42, 44–5,
    53
messages, post-humous 182–7, 189–99,
    214–15, 216–22, 231
    continuation-type 192–5, 197–8
    video 197–8
millennials 199
mindware operating systems 182
misinformation 26
MIT Bootcamp 196
mobility 9
Monkhouse, Bob 217–18, 231
moral responsibility 99–100, 153–4,
    158–9
Mount Auburn Cemetery 10
Mount Rainer Cemetery 160
mourning
    'care culture' of 42, 44–5
    communities of mourners 65
    disenfranchised mourners 65
    'memory culture' of 42, 44–5, 53
    on social media 1, 2
    see also grief
mummies, right to privacy 114–15
murder 70–81, 99, 157
    perpetrators 76–7
musicians, deceased, holographic
    projections of 204–8, 218
My Farewell Note 189–93, 201, 214, 219
My Wishes 219
MySpace 49, 140

Narrative clips 17, 171
National Public Radio (NPR) 161
'natural persons' 85–6, 115, 117
Necropolis Railway 10
Netflix 17, 177
New Age thinking 55
New England settlers 9
New York Magazine 177
New York Times (newspaper) 26, 32, 110
next of kin 90, 92, 93, 99

and access to digital remains 119, 122,
    124, 149–51
and control of digital remains 155–7
and deletion of Facebook pages 149–51
and Legacy Contacts 96
    see also families; parents
Nicolson, Rosa 66
Nicolson, Vanessa 66, 103
NME 206
Noble, Jason 220–6, 228–30
Norris, James 19–20, 22, 188, 215–19
Northern Ireland 189
Norway 137
NPR see National Public Radio
nutrition 9

open-source software 175
openness 16
Orbison, Alex 206, 207
Orbison, Roy 203–4, 206
Otsuchi 6–7
Overcoming Obstacles (blog) 212
Oxford Internet Institute 60

paedophiles 238
Panel on Death and Post-Mortem Privacy
    81–2
paper records 91
parasocial relationships 224, 238
parents
    and access to children's personal online
        data 28–9, 90–2, 102–7, 120–7, 127–8
    and control of Facebook sites 65–7
    and sharenting 235–9
    and their children's privacy rights 235
    see also families; next of kin
partners
    and personal data access 126–7
    see also spouses
passwords 21, 22
    master-password systems 246–7
Peppers ghost projection trick 205
Perpetu 181
perpetuity 180–1
    assurances of 183
    of cemeteries 164–6
    and holographic projections 204
    myths of 227
personal data
    accessing the dead's 28–9, 63–7, 80,
        90–2, 94–7, 102–8, 119–28, 131–2,
        156, 197
    amounts generated 166
    categories of 117

ease of handing over 110
individual wishes regarding following
    death 126–7
and the law/informational privacy 85,
    109
and privacy of the dead 117
storage crises 166
see also hyper-personal data
personally identifiable information 117,
    127
personhood, transcendence of death 118
pet cemeteries, virtual 46–7
Pew Research Center 76
Pezzuti, Jeff 206, 207
phonography 5
photo-montage memorials 170
photographs 93–4, 248
    digital archives of the deceased 145
    physical 228
    spirit 3, 5
    upsetting 76–8, 80, 92, 103–4, 106, 157
Pinterest 52
Pitsillides, Stacey 181, 183
Plug, Sander 31
porn, revenge 195
possessions of the dead
    disaster victims 48
    sorting through xi–xix, **xii**
Presley, Elvis 203
Prince 224
privacy xix–xx, 111–32, **112**
    bodily 109, 117
    and context collapse 16
    demise of 109–11
    and digital natives xxii, 16
    dignitary 112–13
    and Facebook 92–3, 109–11
    group 127–32
    informational 109, 116–17, 127
    laws 89, 109, 131
    legal right to 101, 111, 114–15, 116–32,
        234–9
    and life loggers 17
    meaning of 108–9
    resource 112–13
    territorial 109, 111, 117
privacy behaviour 111
privacy concerns 111
privacy of the dead 28–9, 113–27, 118,
    126, 131–2, 234
    and Facebook 50, 75, 100, 101
    right to 101
    and stewarding of digital archives
        148–9

privacy paradox 111
privacy policies 83
private information 110
private–public boundaries 110
probate 22
prostate cancer 217–18
Protestants 54
P.S. I Love You (2007) 184–5, 187, 192
Psychological Society of Ireland 34, 35–6
public information 110
public–private boundaries 110

QR codes 161–2, 163, 164
Quiring, David 60
Quiring Monuments, Inc., Seattle 159–62,
    164

Reece, Jon 159–62, 163, 164
religious affiliation, decline 54–5
Remembering Jason Noble Facebook
    memorial group 224–5
res cogitans (being of mind) 53
res extensa (being of body) 53
resource privacy 112–13
respect for the dead 114
Rest in Pixels (documentary) 214, 215
resurrection 150
revenge porn 195
Revised Uniform Fiduciary Access to
    Digital Assets Act 2017 156
Rickman, Alan 224
rights 108–9, 111
    of the dead 85–6
    to digital assets 90, 91–2, 96–9
    intellectual property 118, 206
    privacy 101, 111, 114–15, 116–32,
        234–9
road traffic accidents 145–6
robots 182, 192
Rolling Stone (magazine) 222
Rose Theatre, New York 204, 206
Rothblatt, Martine 182
Rotten Tomatoes 172
Rycroft, Gary 78–9, 84, 86–7, 98

SafeBeyond 185
Samaritans 191
Saner, Marc 46, 162–3, 227
Sasaki, Itaru 5–6, 57
science fiction 167–78, 180–2, 186, 202
Scientific American (magazine) 5
screengrabbing 225–6, 228–9
séances 5, 205
searching-and-calling reflex 6

Seattle 160
'second death' 125
Second Life 46–7, 202
Second World War 5
self, digital 64
sensitive identifiable information 117, 127
sharenting 235–9
Shavit, Vered 32, 179, 185, 186, 219
Silicon Valley 8, 188–9
smartphones 13
    filming murders on 70–1
    and hybrid cemeteries 161
    and surveillance technology 30
    unlocking the dead's 22
SMS text messaging 15, 37
social media
    and grief 37–8, 49–50
    wills 119
    see also specific social media platforms
soldiers 27–9, 90–2, 127–8, 130, 183–4,
    189–90, 193
Sonos 232
Spain 110
spirit communications 4–5
spirit photography 3, 5
Spiritualist movement 5, 205
Spotify 52, 232
spouses
    and personal data access 126–7
    see also partners
stalkers, deceased 194
stalking by proxy 186–7
Stardust Symposium 1
Stevenson, Robert Louis 69
storage systems 226–9
    fragility of digital 64, 226–7, 236, 249
    hard copy 91, 236–7, 249–50
succession, laws of 19, 89, 90, 93, 118
suicidal persons, and post-humous
    messaging services 191, 195
suicide, completed 37–8, 103–8, 119–21,
    129, 249
surveillance technology 29–30, 109, 238
Swank, Hilary 184
Sweden 55, 137
SXSW 188–9, 215

taboos 8–9, 197
talking about death 245
talking to the dead 2, 5–7, 51–4, 56–8
Tao of Zoe, The 237
TechCrunch 109
Telegram app 175
terminal illness 193, 208–15

terms and conditions 27–8, 82–8, 90, 99,
    111, 128, 244, 246
territorial privacy 109, 111, 117
Time magazine 14
    'Man/Person of the Year' editions
    11–12, 13
Tōhoku tsunami 6
Toolis, Kevin, *My Father's Wake: How the
    Irish Teach Us to Live, Love and Die*
    (2017) 36–7
Toolis, Sonny 36–7
transport 9
trolling 71–3, 81, 116, 118
Troyer, John 22, 60–1, 97, 107, 227
Tumblr 150
Tutankhamun 108, 114, 132
Twain, Mark 26
Twitter 52–3, 77, 139–41, 189, 213, 218
    birth 13
    death notifications 67

'uncanny valley' 176, 181, 187, 197, 205
Uniform Fiduciary Access to Digital
    Assets Act (UFADAA) 97
United Kingdom
    lack of recognition of electronic wills
    96–7
    post-mortem privacy laws 131
United Nations 60
    Universal Declaration of Human Rights
    108–9, 111
United States
    and digital wills 97
    and the law 88–9
    and religious affiliation 54
'Unquiet Grave, The' (folk song) 229–30
urban burial sites, overcrowding 10
urbanisation 10
Ursache, Marius 167–9, 179, 196–202, 219
USA Today (newspaper) 161
user licenses 19

V&A Museum, London 178–9, 196
vaccination 9
View, The (TV show) 182, 192
Vimeo 27, 170
Virginia Tech massacre 2007 153–4
virtual cemeteries 45–7, 53, 61, 138,
    200
    Facebook as 60, 159
    fees 47–8
    pet 46–7
vlogs (video blogs) 15
voice

recognisability 203
  voices of the dead 5, 203–31
'voice from beyond the grave' phenomena
  avoidance 244–5
  and hackers 186

wakes 36–7
Walter, Tony 42, 55, 56, 126
Warwick University 181
Washington Post (newspaper) 205
water sanitation 9
Watson, Emma 17
Watts, Lucy 207–16, 219–20
Web 2.0 12–13, 15, 44
web presence, non-existent/ill-tended
  13–14
Web Sheriff 79–80, 92
websites, memorial 140
WhatsApp 235
White, Geoff 30

'Who Wants to Live Forever?' exhibition,
  London, 2018 178–9, 196
Wilkinson, Glenn 30
Williams, Robin 173
wills 21, 89–90, 96–7, 98, 119, 195, 246
  digital 89–90, 96–7, 246
  social media 119
wind telephone (*kaze no denwa*) 5–7, 57
Wizard of Oz moments 176
World Wide Cemetery 26, 45–7, 50, 60,
  138, 162–3, 180, 227

Yahoo! 27–9, 90, 127–8, 225
YouTube 13, 26, 100, 216–17, 232, 233
  and photo-montage memorials 170
  and trolling 116

Zoe Dialogues 236–7
Zuckerberg, Mark 13, 49, 58, 109–11, 128,
  151